Building and Using a Groundwater Database

Building and Using a Groundwater Database

Garry W. Rowe
Sylvia J. Dulaney

LEWIS PUBLISHERS

Library of Congress Cataloging-in-Publication Data

Building and Using a Groundwater Database / au. Rowe, Garry W. and Dulaney, Sylvia, J.

Bibliography : p.
Includes index.
1. Groundwater — Database. 2. Well construction.
I. Rowe, Garry W. II. Dulaney, Sylvia, J.
GB428.W97 1991 91-76319
ISBN 0-87371-404-0 CIP

LEWIS PUBLISHERS, INC.
121 South Main Street, P.O Drawer 519, Chelsea, Michigan 48118

PRINTED IN THE UNITED STATES OF AMERICA

Preface

This book is a discussion of the development of a groundwater analysis tool, the groundwater database, but does not provide training or education in the fields of geology, hydrogeology, soil science, water chemistry, water well drilling, cartography, or computer science. Many books and courses are already available to anyone pursuing a certain level of expertise in any one of these fields. Developing and using a groundwater database, however, does involve a certain level of expertise in any one or more of these fields of study, as well as new skills not previously associated with these fields of study (e.g., public relations, administration, management, writing, computer programming).

A groundwater database is herein defined as a computer database management program that includes both well construction and water chemistry information or data. This program should be capable of interfacing with other software programs and groundwater models to analyze hydrogeologic conditions for a selected area.

Analysis of hydrogeologic conditions is necessary for an adequate assessment of a site for land use planning, zoning restrictions, and the cleanup of actual pollution problems. With the ever-increasing volume of data and information used from and for hydrogeologic studies, a tool is needed to manage, manipulate, and coordinate it. The groundwater database is the most effective means to accomplish this.

The task of creating a groundwater database is monumental, but it should prove to be a time-saving tool which, in turn, will improve the quality of the final product. The goal of this book is to introduce the individual, office, or agency to basic principles involved with groundwater database development.

Four distinct activities are involved in the development of a groundwater database: (1) development of the well construction or well-log database, which includes the mapping of well locations, (2) use of the well construction database to describe hydrogeologic conditions for a study area, (3) sampling of selected wells in a study area for water chemistry, and (4) development of a water chemistry database. Subsequent examples show how the groundwater database can be used.

Garry W. Rowe is a Program Specialist with the Ingham County Health Department, Division of Environmental Health, in Lansing, Michigan. Mr. Rowe works in the Program of Toxic Materials and Hazardous Waste, and is responsible for several programs associated with groundwater protection and monitoring. This includes the development and maintenance of a computerized groundwater database of water well records and water chemistry.

Mr. Rowe has been involved with a variety of state and local programs associated with groundwater database development and groundwater protection. This includes chairperson for the Technical Core Group for the Southwest Michigan Groundwater Survey and Monitoring Project, Chairperson for the Groundwater Technical Advisory Council of the Groundwater Management Board for the Tri-County Regional Area of Lansing, Michigan, and a member of the Statewide Groundwater Data Base Technical Advisory Board.

His academic studies include a Bachelor of Science in Agriculture from the University of Arizona and a Masters of Science in Crops and Soil Science from Michigan State University. Prior to working with the Ingham County Health Department, Mr. Rowe was employed with the Ingham County Soil Conservation District as a Conservation Technician.

Sylvia Dulaney is co-founder of Pass.word, a consulting company specializing in software and training for environmental professionals. As consultant to the Michigan Groundwater Survey (MGS), she trained health department sanitarians, geology students, and resource managers to use microcomputers for groundwater work. Pass.word writes software for environmental professionals, including MGS software for water wells and groundwater chemistry, Aquipro software, and data conversion software.

Environmental publications include GIANT STEPS, a technical newsletter for microcomputer users, and papers on technology transfer at users' conferences. Prior to environmental specialties, Sylvia authored movies, books, and articles on organization and learning.

Current activities focus on integrating geographic information systems into local planning. She consults with county governments, schools, health departments, and planning agencies on ways to implement map technology into existing organizational structures.

Acknowledgments

We wish to express our sincere thanks to members and staff of the Michigan Groundwater Survey who originated many of the procedures and techniques described in this book. In addition, several illustrations are due to resources provided by the Survey such as software programs, water tests, and training in groundwater databases and computers. M.G.S. activities were hosted by Western Michigan University's Science for Citizens Center, and were funded with the generosity and foresight of The W. K. Kellogg Foundation.

We also thank the staff of the Center for Remote Sensing, Michigan State University, who developed CMap to include groundwater users, and the Michigan Resource Inventory System (MIRIS) in Michigan's Department of Natural Resources, who digitized Michigan's resources and base maps.

The authors thank the Statewide Groundwater Data Base Project in Michigan's Department of Natural Resources for all the work and development that went into creating an on-going statewide groundwater database system. This work was a valuable insight into all the specific task problems encountered when developing such a project.

Appreciation is also extended to the Ingham County Health Department for the use of case examples and data which were invaluable to the examples portrayed throughout this book.

We thank the people and organizations who agreed to let us present their material in this book. Their efforts and generosity are deeply appreciated, and we praise them for the contributions they make protecting our groundwater resources.

Table of Contents

CHAPTER 1

Getting Started – Why and What Size

1. GROUNDWATER DATABASES AND GROUNDWATER PROTECTION

Groundwater management involves the coordination of data, policy, and informed persons. Those in a position to make and enforce law and policy for protecting groundwater resources need consult with those who have access to groundwater data and can properly interpret it. Interpretation and consultation are only as good as the available data and the trained people using it. Even if all the information is available, it still takes an individual with management skills to analyze the data properly and provide accurate interpretation for groundwater management decisions.

There are many components to groundwater data, including groundwater flow direction, water quantity, water quality, lithology of bedrock, and lithology of glacial drift. Groundwater data are generally used to investigate or prevent a known pollution problem. Most cases are investigatory, conducted after an individual or a company receives a written citation for a pollution incident.

Part of the reason for this crisis-reaction as opposed to a planning-prevention stance is the attitude of "only spending the money if I really have to." It is a well-established fact that hydrogeologic work is not inexpensive; however, it can usually be demonstrated that "an ounce of prevention is worth a pound of cure," if the ounce is paid for up front.

Collecting and organizing current data for any hydrogeologic study entails significant time and effort. The investigatory process involves traveling to agencies or companies, pulling files, sorting material, and conforming data into an organized system for study. After organization comes manipulation and sometimes reentering other programs for analysis. Accomplishing this manually is tedious and very time consuming, and is usually a susbstantial strain on budgets. When one looks at the costs of this process, one can easily understand the "only spending the money if I really have to" attitude.

2. AN ELECTRONIC SOLUTION

Systems are needed to provide data conveniently and to manipulate information for analysis. Groundwater data should be computerized so that information be accessible to a variety of users and communicated across agencies and programs at the federal, state, and local levels. Until the groundwater investigative process can be made more efficient, groundwater pollution, prevention, and management may never be effectively developed.

Various agencies realize the need for a computerized groundwater database to handle an ever-increasing amount of information. In September 1984, the State of Michigan published "Groundwater Protection Initiatives,"[1] which listed several programs needed for protecting groundwater. Recommendation 16 specifically called for the State to develop and implement a coordinated program to computerize all groundwater data for quick storage, handling, and retrieval. The program should include information that would help businesses select sites that would not impair usable aquifers and be useful and accessible to local agencies.

In March 1983, several agencies in Oklahoma recognized the need for an effective groundwater management program to address the increasing problems of pollution, quantity, land subsidence, and saltwater intrusion.[2] In outlining their strategy, the first objective of the plan was to define the physical characteristics of the groundwater system. To accomplish this they established a goal: "to develop a computerized data base which will assist in evaluation of the local ground water system." Databases need to be assembled and computerized so information can be easily accessed and useful.

The U.S. Environmental Protection Agency (EPA) Office of Ground-Water Protection defines the structure of a system as one that will "store, share, and manipulate ground-water data to support better environmental decision-making...."[3] Based on interviews, workshops, and a review of the literature and existing systems, requirements and standards for a groundwater data management system were studied. It was found that decision makers required improved capabilities in three areas: data access, data retrieval, and data analysis.

In 1980, the U.S. Congress passed the Comprehensive Environmental Response, Compensation and Liability Act (CERCLA). This law, commonly referred to as Superfund, is designed to provide funding for the cleanup of hazardous waste sites throughout the U.S. A key component to this Act is that it allows the EPA to act "immediately" on a site if human health and the environment are threatened, and later recover the cost from the responsible party. With the creation of the Act, the EPA has the primary responsibility for managing cleanup and enforcement of activities associated with any site. The EPA will enter into an agreement with state agencies to coordinate efforts to address each site and require each state to pay a certain percentage of the cleanup costs. As expected, a task of this size requires the cooperation of the EPA, local agencies, and private contractors. Private contractors and consultants will usually perform most of the on-site work.

Because of the possible involvement of several groups and the need to expedite action, databases of information on these sites are necessary to help disseminate data to each group and avoid the legendary overlap of effort; i.e., too often each group may independently collect the same data, make the same types of maps and diagrams, and prepare the same reports as another group, wasting more time and money. If the groups involved take the extra time to computerize their work, the existing data can provide an ongoing format for information sharing.

Chapter 11 discusses the use of a database to analyze a Superfund site. Once data are added to the database, they can be used continuously in different ways to review a site. Since the state agencies are usually involved with most of the Superfund sites, the creation of a statewide database of site information would seem the most appropriate place to start a program. Appendix A provides a list of state Superfund offices by EPA region.

3. THE BUDGET AND THE BUREAUCRACY

Decision and policy makers are found along the entire agency-bureaucracy ladder on local, state, and federal levels. The ideal database program needs to address the problem of collecting data and sharing it across all levels of government. To accomplish such a task would mean the establishment of standards for database information and computer tools so that members of the groundwater community are able to communicate. The EPA recognized this and listed the following as the goals of the 1987 Workshop to Recommend A Minimum Set Of Data Elements For Groundwater:[4]

1. To achieve consensus on a minimum set of data elements that would facilitate the collection and sharing of groundwater and related data across agencies.
2. To identify implementation issues that must be resolved to encourage collection of a minimum set of data elements throughout the groundwater community.

General consensus exists among users of groundwater information that data need to move across all levels of government. Figure 1 represents a framework proposed by the EPA for making groundwater information accessible to different participants of the groundwater management community. Local participants were added to the decision maker component since local agencies have increased their involvement, and in some cases are the lead agency for investigating groundwater problem sites.

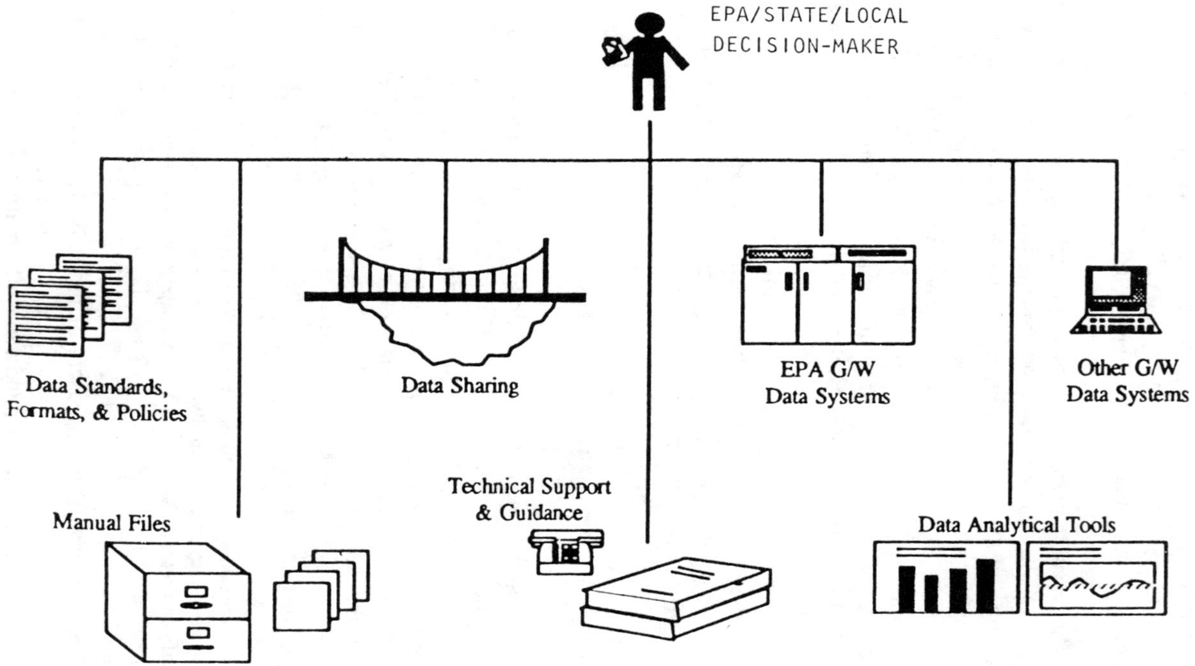

Figure 1. A framework for managing groundwater data and making it accessible to all levels of agency decision makers. (From U.S. EPA Office of Groundwater Protection.)

While it is agreed that a database system is needed, who will develop it? Someone or some organization must take the initiative. Since database needs are found along several levels of government, some coordinated effort is needed across all levels. The EPA has reviewed several scenarios, ranging from a posture of no new initiatives to requiring standardized programs on the state level which are coordinated and funded by the EPA. The more aggressive the program, the more expensive it becomes. While a standardized program initiated on a federal level may address a generalized need for the entire country, it may not meet specific needs for local agencies. Human nature wants to exercise some control regarding that which has significance in life. What may be best is a program initiated on the federal level with some standardization of data, which would entail state management and local construction. Local agencies would have some control as to the type of data collected so that specific needs central to that local area are met. The database program basically would be standardized to accommodate the need for communication between the different levels of government, but flexible enough to accommodate local needs.

The success of a program depends on the attitude of the people involved and its importance to them. Until the decision makers understand the necessity of a groundwater database to properly manage a groundwater pollution prevention program, individuals in the water management industry will be at the mercy of the crisis-reaction approach to solving problems.

4. GETTING STARTED AT A LOCAL LEVEL: THE POLITICS OF SCALE

When a groundwater database is used for planning and decision making, it becomes a geographic information system (GIS). The ideal database unit or GIS has the following characteristics.

The GIS is organized around political/decision maker boundaries. Databases and related geographic information systems are useful as decision support, and should be tied to decision makers. One can argue that a GIS does not exist except within the context of a political homeland. The database should serve a unique local self-governing body such as a township or county. Two queries may be posed: "does this boundary make sense to the people we intend to serve?," and, "is this boundary feasible given the expertise of our team?"

The size can be supported by available hardware, software, data, timetable, and budget. Databases must be built, and that takes time, patience, and equipment, three scarce qualities during startup. When building a database, keep in mind that many different personalities are necessary, and that the building process is a vehicle to develop working relationships with others. The initial phases of a project often set the mood for months to come, so start with pilot projects, and pay special attention to quality and to humor.

The Working Group

Several specialties are involved in building and using a groundwater database. In the process, the team will require an assortment of people who excel at the following jobs:

Well Evaluator: Someone who reviews the written well logs to evaluate their suitability for inclusion.

Locator: A person who knows the local area and can locate wells accurately on municipal or topographic maps.

Digitizer: Someone who can operate a computer and digitizing software. It is best to let one person do the well digitizing for several locators as it is a dizzying task.

Database Manager: Someone who can program well enough to manage the database as it is built. This person gives support to the people who enter the well records and writes programs to produce special reports or results.

Hardware Manager: One who can keep plotters, digitizers, modems, printers, and mice working. This person should maintain files on equipment, manuals, proofs of purchase, and warranties.

Keyboard Operator: Fast, accurate typists who use database software to enter well construction data. Keyboard entry is aided by software features like error checking and function keys.

Database Quality Control and Inventory: This person checks the completed electronic database for accuracy — is the well located correctly, is the well log text correct? As database librarian, this individual inventories and catalogs existing databases and how they are used, and is ultimately responsible for data backups and procedures.

User: This person is the "end user," someone who might say, "I need a map showing shallow wells within 2 mi of this site." The database project is more effective when active users on the project and from other agencies are involved.

Hydrogeologist: Interprets groundwater data and geologic information. Their input during well selection, sampling, and aquifer characterization saves time and makes the greatest use of available data.

Writer: One who assembles the maps, reports, and tables into a coherent report. He/she may be the group's expert on word processing or desktop publishing.

Graphics Manager: How the graphics are prepared can vary — slides can be produced from computer plotter files, cameras can take pictures of the screen, etc. This person needs an eye for cartography and the ability to communicate.

Each of these specialties requires some mix of local knowledge and computer skills. On a scale, each job has a different ideal mix:

	Local Expert					Computer Expert
Well evaluator		X				
Locator	X					
Digitizer				X		
Database manager						X
Hardware manager						X
Keyboard operator			X			
Data quality control and inventory					X	
User	X					
Hydrogeologist			X			
Writer		X				
Graphics manager					X	

Establishing Goals

The way to establish goals is to collect data on people's understanding of the project at hand and then begin to assemble a prototype of the finished product. The easiest goals to establish are tangible. For example, "Add 100 New Wells to the Database in Atlas Township," or "Produce A Topographic Map of Bedrock Elevations in Barry County." Other goals are less tangible: "Prepare Useful Information for Land Use Plan Revision." Make goals as tangible as possible so all members of the team can understand what the finished product will look like. This is accomplished by consistently asking "how:"

"Prepare Groundwater Information for Land Use Plan Revision"

"How?"

"By showing maps of groundwater-sensitive areas, contaminated areas, protected areas."

"How?"

"Get previous reports, compile relevant information to one working map."
"Gather well logs of selected areas."
"Plot cross-section of wells to show sensitive groundwater areas."

Soon, a list of tasks will be assembled, and the team can be organized around the tasks and a timetable.

REFERENCES

1. State of Michigan, Groundwater Protection Initiatives, Committee of the Cabinet Council on Environmental Protection (1984).
2. Association of Central Oklahoma Governments Garber-Wellington Association and the EPA's Robert S. Kerr Environmental Research Laboratory, Executive Summary for Establishment of a Ground Water Management and Protection Plan. Oklahoma City (March 1983).
3. Ground-Water Data Requirements Analysis, EPA 440/6-87-005, U.S. EPA Office of Ground-Water Protection, Office of Information Resources Management (May 1987).
4. EPA Workshop to Recommend A Minimum Set of Data Elements for Groundwater Water, Workshop Findings Report, EPA 440/6-88-005, USEPA Office of Ground-Water Protection (June 1988), p. 1.

CHAPTER 2

Database Backbone – Information from the Well-Log

1. THE WELL CONSTRUCTION RECORD

Every system, organic or inorganic, needs a foundation. For houses it is concrete and 2 × 4s; for human beings it is the skeleton; and for a groundwater database, it is the well driller's well construction record or well-log. This individual document supplies all historic evidence that a well exists, from which groundwater and information on geologic material can be obtained.

The availability of well construction records or well-logs varies throughout the U.S., and the extent of the data included in each record may differ from state to state. The National Water Well Association in Dublin, OH, maintains a status list of well construction or "well completion reports" for the U.S.[1] With a few exceptions, all states currently have a program that addresses the submission of well construction reports by drillers of wells used for drinking water.

A 1989 report by Mary C. Ganley[2] gives extensive information on well construction record availability throughout the U.S., including an evaluation of the general quality of the data. Table 1, which came from that report, gives information on the agency holding records in each state. Table 2, also from that report, outlines the types of information available in the record for each state.

Programs for collecting well water records are usually handled at the state level and it is the responsibility of the individual state to develop its own program; each state also determines what should be included on the well construction form. Examples of some well construction record forms are given in Figures 1 to 4. Appendix B lists state agencies for obtaining information on well construction records.[3]

Table 1. Domestic Well Record Availability in the United States

State	Legal Requirement Year	Agency and City Location	Hardcopy Record Filed by	Database	Comprehensive for All Counties	Other Record Locations
1 Alabama	1972	AL Geological Survey, State Oil and Gas Board, Tuscaloosa	County, grid system	Yes	Yes	Some local county health depts.
2 Alaska	1972	AK Dept. of Natural Resources, Div. of Geological & Geophysical Surveys, Eagle River	TRS or index number if unlocatable	Yes	Yes	Borough Depts. of Environmental Conservation
3 Arizona	1980	AZ Dept. of Water Resources, Phoenix	TRS	Yes	Yes	None
4 Arkansas	1969	AR Comm. on Water Well Construction, Little Rock	Contractor, year	Working on	Yes	AR Geological Commission, University of Arkansas, School of Engineering
5 California	1951	CA Dept. of Water Resources, Red Bluff, Sacramento, Fresno, Los Angeles	TRS	Not statewide	No	Four separate regions
6 Colorado	1972	CO Dept. of Natural Resources, Div. of Water Resources, Denver	Permit #	Working on	Yes	Duplicates in seven regional offices
7 Connecticut	1955	CT Dept. of Environmental Protection, Natural Resources Center, Hartford	Town, construction date	Yes	Yes	Conservation Commission, various county agencies
8 Delaware	1969	DE Dept. of Natural Resources and Environmental Control, Water Supply Branch, Dover	Permit #, cross-referenced to owner's name	Yes	Yes	Selected records at and share database with DE Geological Survey
9 Florida	1980	A. St. Johns River Water Management District, Palatka	Permit #	Yes	Yes	No
		B. South Florida Water Management District, Miami	TRS	Yes	Yes	Some counties
		C. Northwest Florida Water Management District, Havana	Permit #, year	Yes	Yes	No
		D. Southwest Florida Water Management District, Brooksville	Permit #	Yes	Yes	No
		E. Suwannee Water Management District, Live Oak	Permit # or TRS	Yes	Yes	No
10 Georgia			No legal requirement			
11 Hawaii	1988	HI Dept. of Land and Natural Resources, Div. of Water and Land Development, Honolulu	Well number (assigned based on lat-long)	None	Yes	Dept. of Health
12 Idaho	1967	ID Dept. of Water Resources, Boise	TRS	None	Yes	Copies in four regional offices
13 Illinois	1967	A. IL State Geological Survey, Champaign	County, TRS	Yes	Yes	IL Dept. of Public Health, county agencies
		B. IL State Water Survey, Champaign	TRS or city	None	Yes	IL Dept. of Public Health, county agencies
14 Indiana	1959	IN Dept. of Natural Resources, Div. of Water, Indianapolis	TRS	No	Yes	County agencies
15 Iowa			No legal requirement			
16 Kansas	1974	KS Dept. of Health and Environment, Bureau of Water Protection, Topeka	County, TRS	Yes	Yes	KS Geological Survey
17 Kentucky	1985	KY Natural Resources and Environmental Protection Cabinet, Frankfort	USGS Quadrangle, aqua number	Yes	Yes	KY Geological Survey, Cabinet for Human Resources, county health depts.
18 Louisiana	1985	LA Dept. of Transportation and Development, Office of Public Works, Water Resources Section, Baton Rouge	Parish, well number	Yes	Yes	None
19 Maine	1987	ME Geological Survey, Dept. of Conservation, Augusta	County, chronologically, well number	Yes	Yes	None
20 Maryland	1946	MD Dept. of Health and Mental Hygiene, Baltimore	County, well permit #	Yes	Yes	Yes
21 Massachusetts	1962	MA Dept. of Environmental Management, Div. of Water Resources, Boston	City and town, community	No	Yes	Some local boards of health
22 Michigan	1967	MI Dept. of Natural Resources, Geological Survey Div., Glacial and Groundwater Unit, Lansing	County, TRS	For Isabella Co. only	Yes	County health depts., MI Dept. of Public Health
23 Minnesota	1975	MN Geological Survey, St. Paul	County, TRS	Yes	Yes	MN Dept. of Health, MN Dept. of Natural Resources, MN Soil & Water Conservation Board, county zoning depts. or health services
24 Mississippi	1963	MS Dept. of Natural Resources, Bureau of Land and Water Resources, Jackson	County, date grid	Yes	Yes	None
25 Missouri	1986	MO Dept. of Natural Resources, Div. of Geology and Land Survey, Rolla	County, TRS	Working on	Yes	Counties have water quality, not well construction info.
26 Montana	1973	MT Dept. of Natural Resources and Conservation, Water Rights Bureau, Helena	Water rights #	Yes	Yes	MT Bureau of Mines and Geology
27 Nebraska			No legal requirement			
28 Nevada	1947	NV Dept. of Conservation and Natural Resources, Carson City	TRS, date	In cooperation with USGS since 1984	Yes	Las Vegas Office
29 New Hampshire	1984	NH Dept. of Environmental Services, Water Well Board, Concord	County, town	Inputs into GWSI and NHGIS	Yes	None

TRS = Township or Tier, Range, Section. lat = Latitude. long = Longitude.

State	Legal Requirement Year	Agency and City Location	Hardcopy Record Filed by	Database	Comprehensive for All Counties	Other Record Locations
30 New Jersey	1948	NJ Dept. of Environmental Protection, DWR, Bureau of Water Allocations, Trenton	Grid system	Working on	Yes	Some county health depts.
31 New Mexico	1931	NM State Engineer Office, Santa Fe	Basin, TRS	None	Yes	Duplicates in five district offices
32 New York	None statewide	NY Dept. of Environmental Conservation, Albany	In county reports area	Yes	No	Not applicable Records kept for Delaware River Basin, seven-county area; records in Long Island are kept at the DEC Region 1 office at Stonybrook
33 North Carolina	1967	NC Dept. of Natural Resources and Community Development, Raleigh	1966-1982: county, grid; 1982+: county	Raleigh region only	Yes	Duplicates in some of the seven regional offices
34 North Dakota	1972	ND Board of Water Well Contractors, Bismarck	County, date, TRS	None	Yes	Local health departments
35 Ohio	1948	OH Dept. of Natural Resources, Columbus	County, TRS	None	Yes	OH Dept. of Health has duplicates dating back to 1977
36 Oklahoma	1982	OK Water Resources Board, Groundwater Div., Oklahoma City	County, TRS	Working on	Yes	None
37 Oregon	1955	OR Water Resources Dept., Salem	County, TRS	Planning as of 1987	Yes	Duplicates in 19 local Water Masters Districts
38 Pennsylvania	1956	PA Dept. of Environmental Resources, Bureau of Topographic and Geologic Survey, Harrisburg	County, township	Yes	Yes	None
39 Rhode Island	1972	RI Water Resources Board, Hydrogeology Div., Providence	Town, year	None	Yes	RI Dept. of Health, RI Dept. of Environmental Management
40 South Carolina	1985	A. SC Water Resources Commission, Columbia	County, grid system	Yes	Yes	Has records, some located by grid, prior to 1985
		B. SC Dept. of Health and Environmental Control, Columbia	County	Yes	Yes	Has kept records since 1985
41 South Dakota	1955, 1972 (1972 mandate)	SD Dept. of Water & Natural Resources, Div. of Water Rights, Pierre	County, TRS	No	Yes	SD Geological Survey
42 Tennessee	1963	TN Dept. of Health and Environment, TN Div. of Ground Water Protection, Nashville	County, original owner, date	Yes	Yes	None
43 Texas	1969	TX Water Commission, Austin	County, grid system	No	Yes	TX Water Development Board, underground water conservation districts
44 Utah	1935	UT State Engineer, Div. of Water Rights, Sale Lake City	File number and TRS	Yes	Yes	None
45 Vermont	1966	VT Agency of Natural Resources, DEC, Water Quality Div., Groundwater Mgt. Section, Waterbury	Town	Working on	Yes	None
46 Virginia	1986	VA Water Control Board, Richmond	Sequential Well Number, county	Yes	Yes	Duplicates in six regional offices, local health depts.
47 Washington	1973	WA Dept. of Ecology, Water Resources Div., Remond, Spokane, Olympia, Yakima		TRS	None	No
48 West Virginia	1984	WV Dept of Health, Office of Environmental Health Services, Environmental Engineering Div., Charleston	County, year, permit #	Yes	Yes	Six districts, county health depts.
49 Wisconsin	1936	WI Geological and Natural History Survey, Madison	County, TRS	None	Yes	WI Dept. of Natural Resources
50 Wyoming	1969	WY State Engineers Office, Ground Water Div., Cheyenne	Permit #, cross-files cards for owner name and well location	Yes	Yes	WY Dept. of Environmental Quality, county zoning boards

Table 2. Technical Information Items and Tabulations for 51 Agency Well Record Forms[a]

	STATES																										
									Florida[b]																		
	AL	AK	AZ	AR	CA	CO	CT	DE	A	B	C	D	E	HI	ID	IL	IN	KS	KY	LA[c]	ME	MD	MA	MI	MN	MS	MO
GENERAL																											
1. Owner's Name	•	•	•	•	•	•	•	•	•	•	•	•	•	•	•	•	•	•	•	•	•	•	•	•	•	•	•
2. Driller's Name	•	•	•	•	•	•	•	•	•	•	•	•	•	•	•	•	•	•	•	•	•	•	•	•	•	•	•
3. Registration/License Number	•	•		•	•	•	•		•	•	•	•	•		•	•		•	•	•		•	•	•	•		
4. Permit Number		•	•		•	•	•	•	•	•	•	•	•		•	•			•			•				•	•
5. Construction Date	•	•	•	•	•	•	•	•	•	•		•	•	•	•	•	•	•	•	•	•	•	•	•	•	•	•
6. Well Log Confidentiality					•																						
7. Well Use	•	•		•	•		•	•	•	•	•		•		•	•	•	•	•	•	•		•	•	•	•	•
8. Nearest Septic, Sewer or Possible Contamination Source								•								•		•				•		•	•		
9. Replacement or Abandoned Wells					•			•	•	•	•		•		•			•	•	•		•			•		•
10. Comments/Remarks		•	•	•				•			•			•		•			•	•	•		•	•	•	•	
CONSTRUCTION DETAILS																											
11. Well Service				•	•				•	•	•		•		•			•	•								
12. Drilling Method	•	•	•		•	•	•	•	•	•		•	•	•	•	•	•		•		•		•	•	•		•
13. Drilling Fluid								•		•									•						•		
14. Open Hole	•										•	•						•							•	•	
15. Total Well Depth	•	•	•	•	•	•	•	•	•	•	•		•	•		•	•	•	•	•	•	•		•	•	•	•
16. Borehole Diameter	•			•	•	•	•							•	•	•		•	•						•		
17. Grout	•	•		•	•	•	•	•	•	•	•	•	•	•	•	•		•	•	•		•		•	•		•
18. Grout Material		•		•		•		•		•		•			•	•		•	•	•		•		•	•		•
19. Grout Interval	•			•	•	•		•	•	•	•	•	•	•	•	•		•	•	•		•		•	•		•
20. Grout Method					•	•									•					•		•			•		•
21. Packers	•														•												
22. Gravel Pack	•	•		•	•	•	•	•						•	•	•	•	•	•		•	•	•			•	•
23. Gravel Pack Material								•											•								
24. Gravel Pack Size					•	•	•								•												•
25. Gravel Pack Interval				•	•	•	•	•						•	•			•	•			•					•
26. Drive Shoe							•								•	•								•	•		•
27. Casing Gauge		•		•	•		•							•	•	•		•		•				•	•		•
28. Casing Diameter	•	•	•	•	•	•	•	•	•	•	•	•	•	•	•	•	•	•	•	•		•	•	•	•	•	•
29. Casing Material	•		•	•	•	•		•	•	•	•	•	•		•	•		•	•	•		•	•		•	•	•
30. Casing Interval or Length	•	•	•	•	•	•	•	•	•	•		•	•	•	•	•	•	•	•	•	•	•	•	•	•	•	•
31. Casing Stick Up		•						•								•		•						•	•		
32. Couplings	•	•	•				•								•			•	•					•	•		•
33. Perforated or Screen Material		•	•	•	•	•	•	•		•				•	•			•	•	•		•	•	•	•	•	•
34. Slot or Perforation Size	•	•	•	•	•	•	•			•				•	•	•	•		•	•		•	•	•	•	•	•
35. Perforation or Screen Interval or Length	•	•	•	•	•	•	•	•	•	•		•		•	•	•	•	•	•	•		•	•	•	•	•	•
36. Well Sanitation				•												•		•	•					•	•		•
37. Pump Information	•	•	•			•		•	•	•	•			•		•	•		•			•		•	•	•	•
38. Surface Completion								•							•	•			•					•	•		
HYDROGEOLOGY																											
39. Static Water Level	•	•	•	o	•	•	•	•	•	•	•	•	•	•	•	•	•	•	•	•	•	•	•	•	•	•	•
40. Pumping Water Level	•	•	•		•	•	•	•	•	•	•	•	•	•	•	•	•	•	•			•	•	•	•	•	•
41. Aquifer Media				•		•			•	•		•	•		•	•					•	•	•		•		
42. Geologic Formation																				•	•					•	
43. Driller's Log	•	•	•	•	•	•	•	•	•	•		•	•	•	•	•	•	•	•	•	•	•	•	•	•	•	•
44. Well Test	•	•	•		•	•	•	•	•	•		•	•	•	•	•	•	•	•				•		•	•	•
45. Estimated Yield	•	•	•	•	•	•	•	•	•	•		•	•	•	•	•	•	•	•		•	•	•	•	•	•	•
46. Water Quality	•				•					•		•		•	•	•	•	•	•		•		•	•			

LOCATION																											
47. Owner Mailing Address	•	•	•	•	•		•		•										•		•		•	•	•	•	
48. Well Location Address		•		•	•	•	•		•	•	•		•		•	•	•	•	•	•		•	•	•	•		•
49. Well Location TRS	•	•	•	•	•	•			•	•	•	•	•		•	•	•	•						•	•	•	•
50. Well Location Lat. and Long.		•							•	•		•	•	•					•				•				
51. Other Location	•	•	•	•	•		•	•	•	•			•	•	•		•	•	•	•	•	•	•	•	•	•	•
52. Well Elevation			•	•										•		•	•	•	•	•				•	•	•	•

NOTE: Various factors may contribute to what are apparent inaccuracies in the data presented in this table. Misinterpretations or errors noticed by readers would be of great interest to the author.

[a]Data are not available for Georgia, Iowa, and Nebraska.

[b]Florida's water management districts are:
A St. John's River
B South Florida
C Northwest Florida
D Southwest Florida
E Suwannee

[c]Louisiana has a separate form for plugging and abandonment.

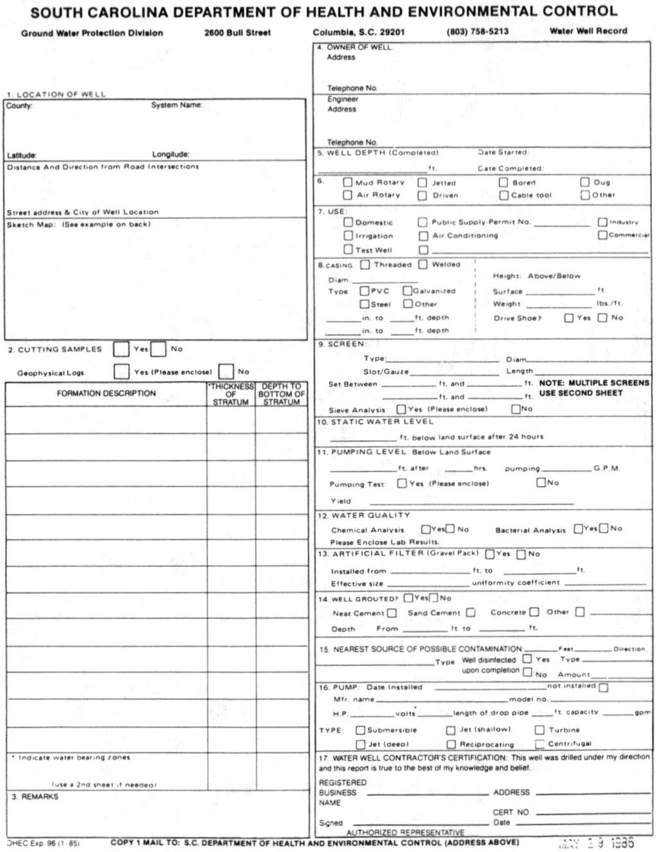

SOUTH CAROLINA DEPARTMENT OF HEALTH AND ENVIRONMENTAL CONTROL

Ground Water Protection Division 2600 Bull Street Columbia, S.C. 29201 (803) 758-5213 Water Well Record

1. LOCATION OF WELL
County: System Name:
Latitude: Longitude:
Distance And Direction from Road Intersections
Street address & City of Well Location
Sketch Map: (See example on back)

2. CUTTING SAMPLES ☐ Yes ☐ No
Geophysical Logs ☐ Yes (Please enclose) ☐ No

FORMATION DESCRIPTION	*THICKNESS OF STRATUM	DEPTH TO BOTTOM OF STRATUM

* Indicate water bearing zones
(use a 2nd sheet if needed)

3. REMARKS

4. OWNER OF WELL
Address
Telephone No.
Engineer
Address
Telephone No.

5. WELL DEPTH (Completed) ______ ft. Date Started: Date Completed:

6. ☐ Mud Rotary ☐ Jetted ☐ Bored ☐ Dug
☐ Air Rotary ☐ Driven ☐ Cable tool ☐ Other

7. USE:
☐ Domestic ☐ Public Supply Permit No. ______ ☐ Industry
☐ Irrigation ☐ Air Conditioning ☐ Commercial
☐ Test Well ☐

8. CASING ☐ Threaded ☐ Welded
Diam. ______ Height: Above/Below
Type ☐ PVC ☐ Galvanized Surface ______ ft.
☐ Steel ☐ Other Weight ______ lbs./ft.
______ in. to ______ ft. depth Drive Shoe? ☐ Yes ☐ No
______ in. to ______ ft. depth

9. SCREEN:
Type: ______ Diam. ______
Slot/Gauze ______ Length ______
Set Between ______ ft. and ______ ft. NOTE: MULTIPLE SCREENS
______ ft. and ______ ft. USE SECOND SHEET
Sieve Analysis ☐ Yes (Please enclose) ☐ No

10. STATIC WATER LEVEL
______ ft. below land surface after 24 hours

11. PUMPING LEVEL Below Land Surface
______ ft. after ______ hrs. pumping ______ G.P.M.
Pumping Test: ☐ Yes (Please enclose) ☐ No
Yield ______

12. WATER QUALITY
Chemical Analysis ☐ Yes ☐ No Bacterial Analysis ☐ Yes ☐ No
Please Enclose Lab Results.

13. ARTIFICIAL FILTER (Gravel Pack) ☐ Yes ☐ No
Installed from ______ ft. to ______ ft.
Effective size ______ uniformity coefficient ______

14. WELL GROUTED? ☐ Yes ☐ No
Neat Cement ☐ Sand Cement ☐ Concrete ☐ Other ☐
Depth From ______ ft. to ______ ft.

15. NEAREST SOURCE OF POSSIBLE CONTAMINATION ______ Feet ______ Direction
______ Type Well disinfected ☐ Yes Type ______
upon completion ☐ No Amount ______

16. PUMP: Date Installed ______ not installed ☐
Mfr. name ______ model no. ______
H.P. ______ volts ______ length of drop pipe ______ ft. capacity ______ gpm
TYPE: ☐ Submersible ☐ Jet (shallow) ☐ Turbine
☐ Jet (deep) ☐ Reciprocating ☐ Centrifugal

17. WATER WELL CONTRACTOR'S CERTIFICATION: This well was drilled under my direction and this report is true to the best of my knowledge and belief.
REGISTERED BUSINESS NAME ______ ADDRESS ______
CERT. NO. ______
Signed ______ Date ______
AUTHORIZED REPRESENTATIVE

DHEC Exp 96 (1-85) COPY 1 MAIL TO: S.C. DEPARTMENT OF HEALTH AND ENVIRONMENTAL CONTROL (ADDRESS ABOVE)

Figure 1. An example of a water well record from South Carolina.

2. WELL CONSTRUCTION DATA: MINIMUM ELEMENTS

A groundwater database must contain certain information to be considered an effective analytical research and investigation tool. In 1987 and 1988, the Ground Water Monitoring and Data Management Office of the U.S. Environmental Protection Agency (EPA) conducted a workshop to determine the most necessary information needed for groundwater data collection. The workshop participants developed a "minimum set of data elements" defined as: "The elements necessary to use data from wells and springs across groundwater related programs." A list of these minimum set components is shown in Table 3. Explanation for these

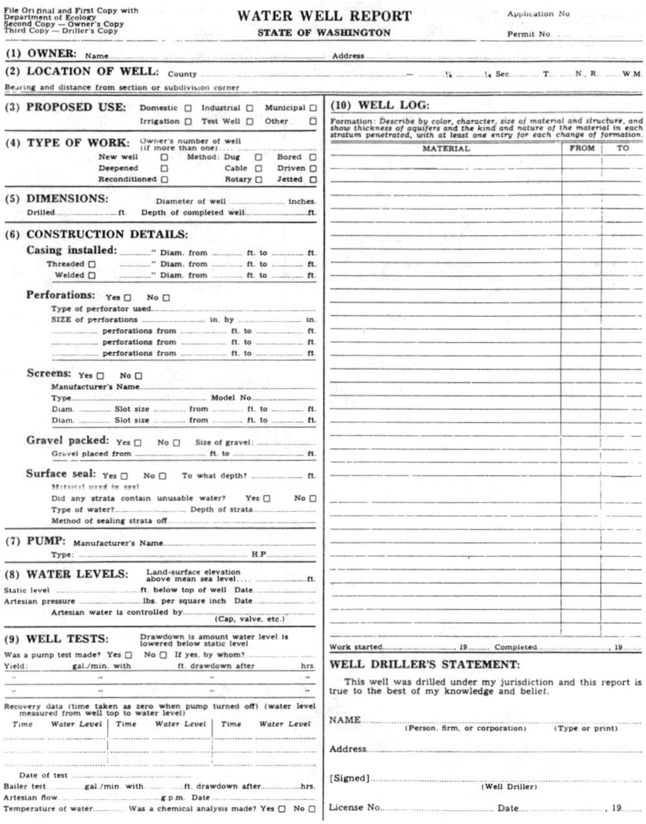

File Original and First Copy with Department of Ecology
Second Copy — Owner's Copy
Third Copy — Driller's Copy

WATER WELL REPORT
STATE OF WASHINGTON

Application No
Permit No

(1) OWNER: Name Address

(2) LOCATION OF WELL: County — ¼ ¼ Sec. T. N., R. W.M.

Bearing and distance from section or subdivision corner

(3) PROPOSED USE: Domestic ☐ Industrial ☐ Municipal ☐ Irrigation ☐ Test Well ☐ Other ☐

(4) TYPE OF WORK: Owner's number of well (if more than one)
New well ☐ Method: Dug ☐ Bored ☐
Deepened ☐ Cable ☐ Driven ☐
Reconditioned ☐ Rotary ☐ Jetted ☐

(5) DIMENSIONS: Diameter of well inches.
Drilled ft. Depth of completed well ft.

(6) CONSTRUCTION DETAILS:

Casing installed: " Diam. from ft. to ft.
Threaded ☐ " Diam. from ft. to ft.
Welded ☐ " Diam. from ft. to ft.

Perforations: Yes ☐ No ☐
Type of perforator used
SIZE of perforations in. by in.
perforations from ft. to ft.
perforations from ft. to ft.
perforations from ft. to ft.

Screens: Yes ☐ No ☐
Manufacturer's Name
Type Model No.
Diam. Slot size from ft. to ft.
Diam. Slot size from ft. to ft.

Gravel packed: Yes ☐ No ☐ Size of gravel:
Gravel placed from ft. to ft.

Surface seal: Yes ☐ No ☐ To what depth? ft.
Material used in seal
Did any strata contain unusable water? Yes ☐ No ☐
Type of water? Depth of strata
Method of sealing strata off

(7) PUMP: Manufacturer's Name
Type: H.P.

(8) WATER LEVELS: Land-surface elevation above mean sea level.... ft.
Static level ft. below top of well Date
Artesian pressure lbs. per square inch Date
Artesian water is controlled by (Cap, valve, etc.)

(9) WELL TESTS: Drawdown is amount water level is lowered below static level
Was a pump test made? Yes ☐ No ☐ If yes, by whom?
Yield: gal./min. with ft. drawdown after hrs.
" " " "
" " " "

Recovery data (time taken as zero when pump turned off) (water level measured from well top to water level)

Time	*Water Level*	*Time*	*Water Level*	*Time*	*Water Level*

Date of test
Bailer test gal./min. with ft. drawdown after hrs.
Artesian flow g.p.m. Date
Temperature of water Was a chemical analysis made? Yes ☐ No ☐

(10) WELL LOG:

Formation: *Describe by color, character, size of material and structure, and show thickness of aquifers and the kind and nature of the material in each stratum penetrated, with at least one entry for each change of formation.*

MATERIAL	FROM	TO

Work started , 19 Completed , 19

WELL DRILLER'S STATEMENT:

This well was drilled under my jurisdiction and this report is true to the best of my knowledge and belief.

NAME (Person, firm, or corporation) (Type or print)

Address

[Signed] (Well Driller)

License No. Date , 19

Figure 2. An example of a water well record from Washington.

elements is given in Appendix C.[4]

The EPA group further defined minimum elements as having the capability of:

1. Communicating groundwater data across programs
2. Being common to all programs, but completely adequate for some programs
3. Providing a road map to other data
4. Providing a link between water quality and well location information

In some areas of the U.S., even these minimum data elements listed by the EPA may not be available, but even if one cannot obtain the minimum element list, a few elements may be satisfactory. If location, well depth, well elevation, and

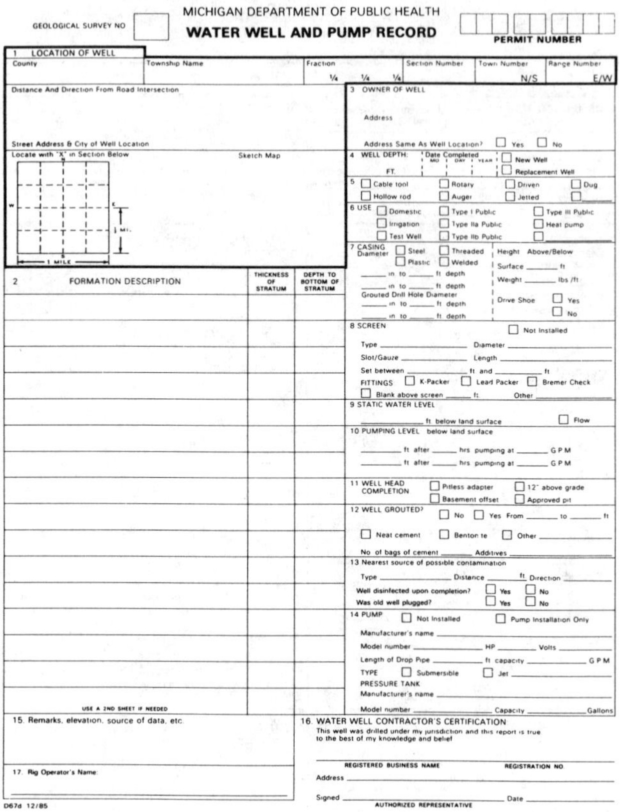

GEOLOGICAL SURVEY NO

MICHIGAN DEPARTMENT OF PUBLIC HEALTH
WATER WELL AND PUMP RECORD

PERMIT NUMBER

1 LOCATION OF WELL
County | Township Name | Fraction ¼ ¼ ¼ | Section Number | Town Number N/S | Range Number E/W

Distance And Direction From Road Intersection

Street Address & City of Well Location

Locate with "X" in Section Below — Sketch Map
N, S, E, W — 1 MILE — 1 MI.

3 OWNER OF WELL
Address
Address Same As Well Location? ☐ Yes ☐ No

4 WELL DEPTH: FT. — Date Completed MO DAY YEAR — ☐ New Well ☐ Replacement Well

5 ☐ Cable tool ☐ Rotary ☐ Driven ☐ Dug ☐ Hollow rod ☐ Auger ☐ Jetted ☐ ____

6 USE ☐ Domestic ☐ Type I Public ☐ Type III Public ☐ Irrigation ☐ Type IIa Public ☐ Heat pump ☐ Test Well ☐ Type IIb Public ☐ ____

7 CASING Diameter ☐ Steel ☐ Threaded ☐ Plastic ☐ Welded
___ in. to ___ ft. depth
___ in. to ___ ft. depth
Grouted Drill Hole Diameter
___ in. to ___ ft. depth
___ in. to ___ ft. depth
Height Above/Below Surface ___ ft
Weight ___ lbs./ft
Drive Shoe ☐ Yes ☐ No

2 FORMATION DESCRIPTION | THICKNESS OF STRATUM | DEPTH TO BOTTOM OF STRATUM

USE A 2ND SHEET IF NEEDED

8 SCREEN ☐ Not Installed
Type ___ Diameter ___
Slot/Gauze ___ Length ___
Set between ___ ft. and ___ ft.
FITTINGS ☐ K-Packer ☐ Lead Packer ☐ Bremer Check
☐ Blank above screen ___ ft. Other ___

9 STATIC WATER LEVEL
___ ft. below land surface ☐ Flow

10 PUMPING LEVEL below land surface
___ ft. after ___ hrs. pumping at ___ G.P.M.
___ ft. after ___ hrs. pumping at ___ G.P.M.

11 WELL HEAD COMPLETION ☐ Pitless adapter ☐ 12" above grade ☐ Basement offset ☐ Approved pit

12 WELL GROUTED? ☐ No ☐ Yes From ___ to ___ ft.
☐ Neat cement ☐ Bentonite ☐ Other ___
No. of bags of cement ___ Additives ___

13 Nearest source of possible contamination
Type ___ Distance ___ ft. Direction ___
Well disinfected upon completion? ☐ Yes ☐ No
Was old well plugged? ☐ Yes ☐ No

14 PUMP ☐ Not Installed ☐ Pump Installation Only
Manufacturer's name ___
Model number ___ HP ___ Volts ___
Length of Drop Pipe ___ ft. capacity ___ G.P.M.
TYPE ☐ Submersible ☐ Jet ___
PRESSURE TANK
Manufacturer's name ___
Model number ___ Capacity ___ Gallons

15. Remarks, elevation, source of data, etc.

17. Rig Operator's Name:

16. WATER WELL CONTRACTOR'S CERTIFICATION
This well was drilled under my jurisdiction and this report is true to the best of my knowledge and belief

REGISTERED BUSINESS NAME REGISTRATION NO.
Address ___
Signed ___ Date ___
AUTHORIZED REPRESENTATIVE

D67d 12/85

Figure 3. An example of a water well record from Michigan.

geologic lithology are available, some level of analysis of a site is possible. Each data user must decide about the data volume, accuracy, and reliability necessary for the job at hand.

3. DATA QUALITY DECIDED AT THE LOCAL LEVEL

Decisions about data quality are usually best addressed at the local or county level where developers of a database know the well driller and the area, and can make a quality control decision about the usefulness of the well-logs they receive.

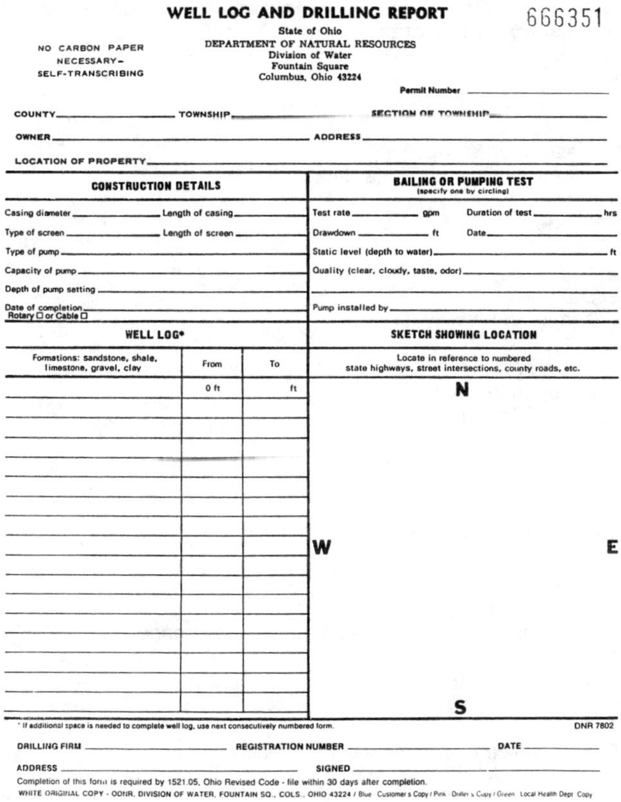

WELL LOG AND DRILLING REPORT 666351

State of Ohio
DEPARTMENT OF NATURAL RESOURCES
Division of Water
Fountain Square
Columbus, Ohio 43224

NO CARBON PAPER NECESSARY– SELF-TRANSCRIBING

Permit Number ______

COUNTY ______ TOWNSHIP ______ SECTION OF TOWNSHIP ______

OWNER ______ ADDRESS ______

LOCATION OF PROPERTY ______

CONSTRUCTION DETAILS	BAILING OR PUMPING TEST (specify one by circling)
Casing diameter ______ Length of casing ______	Test rate ______ gpm Duration of test ______ hrs
Type of screen ______ Length of screen ______	Drawdown ______ ft Date ______
Type of pump ______	Static level (depth to water) ______ ft
Capacity of pump ______	Quality (clear, cloudy, taste, odor) ______
Depth of pump setting ______	
Date of completion ______ Rotary ☐ or Cable ☐	Pump installed by ______

WELL LOG*			SKETCH SHOWING LOCATION
Formations: sandstone, shale, limestone, gravel, clay	From	To	Locate in reference to numbered state highways, street intersections, county roads, etc.
	0 ft	ft	N W E S

* If additional space is needed to complete well log, use next consecutively numbered form. DNR 7802

DRILLING FIRM ______ REGISTRATION NUMBER ______ DATE ______

ADDRESS ______ SIGNED ______

Completion of this form is required by 1521.05, Ohio Revised Code - file within 30 days after completion.

WHITE ORIGINAL COPY - ODNR, DIVISION OF WATER, FOUNTAIN SQ., COLS., OHIO 43224 / Blue Customer's Copy / Pink Driller's Copy / Green Local Health Dept Copy

Figure 4. An example of a water well record from Ohio.

Some interpretation of well driller's jargon may improve the usefulness of the data. For example, some well drillers may mean sandstone when they write rock, and only someone familiar with the area, and drillers may have that awareness.

Some may also consider completing a selection process before including certain well construction records in a database. A data quality standard can be established based on the accuracy of the location, geologic, or construction information. If the person in charge of the selection process is familiar with well drillers in the area, they might also evaluate data quality based on a driller's reputation. Chapter 5 describes several evaluation methods.

Table 3. List of Minimum Set of Data Elements for Groundwater Selected by Workshop Participants

Geographic descriptors
- Latitude
- Longitude
- Method of measure for latitude/longitude
- Source agency for latitude/longitude data
- State FIPS code
- County FIPS code
- Altitude

Well/spring descriptors
- Unique well/spring and facility ID
- Use of well
- Depth of well at completion
- Depth to top of open interval
- Depth to Bottom of open interval
- Location of log
- Type of log

Sample/analysis descriptors
- Source agency for sample data
- Sample date
- Parameter measured
- Concentration/value (in standard units)
- Confidence factor (including field and lab quality assurance)
- Sample ID
- Depth to water
- Measurement quantification

Data from U.S. EPA Office of Groundwater Protection.

REFERENCES

1. Legislative Database, National Water Well Association, 6375 Riverside Drive, Dublin, OH 43017 (June 1988).
2. Ganley, M. C. "Availability and Content of Domestic Well Records in the United States," *Ground Water Monit. Rev.* (Fall 1989).
3. *Computerization of Ground Water Records, Briefings,* Vol. 4, No. 1 (Dublin, OH: National Water Well Association, 1988), p. 7.
4. EPA Workshop to Recommend A Minimum Set of Data Elements for Groundwater Water, Workshop Findings Report, EPA 440/6-88-005, USEPA Office of Ground-Water Protection (June 1988), p. 1.

CHAPTER 3

Methods – Mechanics of Well-Log Mapping

1. LOCATING THE WELL

Groundwater researchers and consultants agree that the single most important aspect of an effective database is having the well construction records located accurately. If the foundation of the database is locationally incorrect, any information for any point within the study area will be misleading. Mapping well construction records is part of the first step toward creating a groundwater database. Accurately locating the well on a map will define the hydrogeology of that site; therefore, one way to select wells to include in a database is whether they can be accurately located on a map.

Work in a few selected counties in the State of Michigan revealed that one third to one half of the well construction records inaccurately listed locations.[1-5] With field verification, up to 77% of the records may be usable.[1,2] The percentage of records that can be successfully located on a map will depend on how tenacious and familiar the database user is with the area.

The degree of accuracy and precision that location data exhibit depends in part on the map scale chosen. For example, a countywide database may be acceptable if wells are accurately located to within 200 ft; however a very site-specific study area may demand accuracy to within 3 ft. The database user must decide.

2. MAPPING SYSTEMS

The land in a number of states is divided or described under the U.S. Public Land Survey System or "Rectangular Survey System," also referred to as the "Tier and Range Survey System." This system became law on May 20, 1785.[6] Public land was surveyed into townships approximately 6 mi^2, and subdivided further into sections approximately 1 mi^2. Two standard lines are used as references, the north-south standard line being the principal meridian, and the east-west standard line being the baseline. "Ranges" of townships parallel to the principal meridian are numbered east and west of this line and "tiers" of townships parallel to the baseline are numbered north and south of this line. Thirty-five different sets of standard lines were established in the U.S. (Figure 1).[6] Townships can be uniquely identified in reference to the tier and range that they occupy (Figure 2).

The most convenient unit used in building a county-sized database is the "Section of a Township." Many well drillers are required to locate a well by county and the tier-range of the townships. Further locational information may include the section of a township and use of a section "quarter" system to delineate the location down to a 10-acre plot (Figure 3). This system is shown in Figure 4 as a well drilled in 1969, located in the SW1/4, SW1/4, NW1/4, Section 21, T.2N., R.1W., of the Vevay Township of Ingham County.

The map most often used to locate wells is the U.S. Quadrangle Map also referred to as a topographical map (or topo). Topographic maps are published by the U.S. Geological Survey and should be available for most locations in the U.S. These maps are convenient for creating a groundwater database because they include many of the data structure items required such as land elevation and three types of geodetic location systems: latitude and longitude, Universal Transverse Mercator, and the State Plane Coordinate system. Figure 5 shows examples of these systems using the Millville Quadrangle.

3. THE MAPPING PROCESS

Because the township is the mapping unit most often used, the township is referred to here as the database unit. For each database unit, the database developer will compile the records available for each individual section. Location information from each well construction record is reviewed to place that record as a point or "dot" on the map.

To facilitate the mapping process, local residents familiar with both the area and citizens, should be involved in the process. Many local health departments have a well permit program whereby sanitarians visit with property owners and/or the well drillers before locating the well. The well and septic system permit may be another document to help pinpoint the well location on the map. Other tools to facilitate the well mapping may include aerial photos, land plat maps, and land surveys. After initial in-house or office mapping is complete, field verification is necessary to confirm the location for questionable records or other records that were impossible to map without a visual field check.

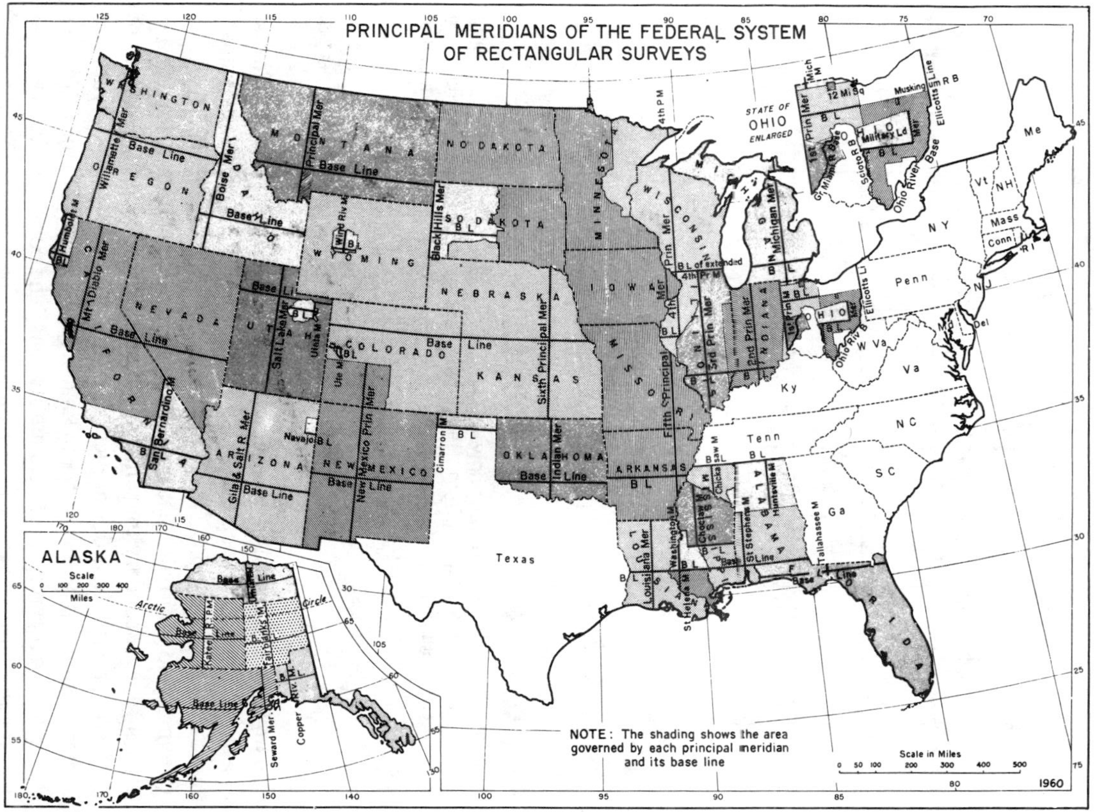

Figure 1. Map of "Standard Lines" for the U.S. Public Land Survey system. (From "Restoration of Lost or Obliterated Corners and Subdivision of Sections: A Guide for Surveyors," U.S. Department of the Interior, 1974.)

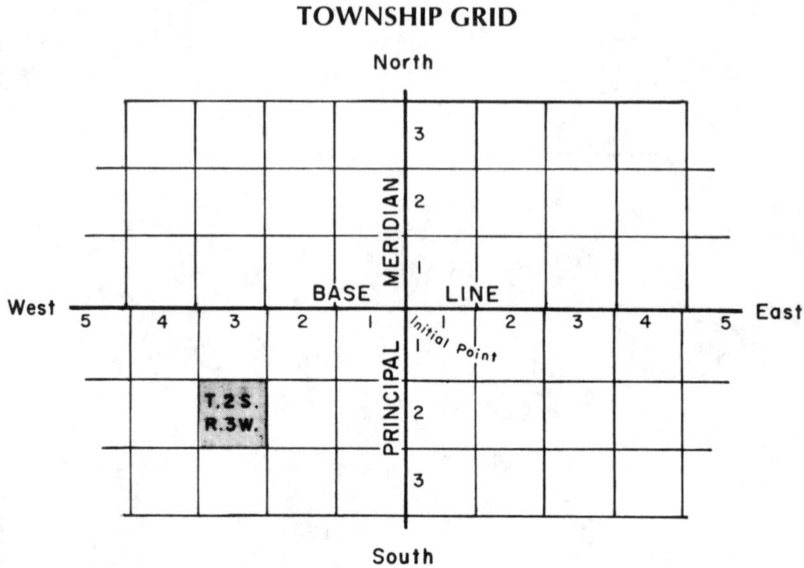

Figure 2.

Figures 2 and 3. Diagrams showing how the U.S. Public Land Survey system is further broken down. (From "Restoration of Lost or Obliterated Corners and Subdivision of Sections: A Guide for Surveyors," U.S. Department of the Interior, 1974.)

As an aid in the field verification process, the database user may want to invest in a field locational tool. One such tool is the Loran-C Radio Navigation System[7] (LongRange Aid to Navigation). The Loran-C is a synchronized pulsed, low-frequency, hyperbolic radio aid that uses a master ground station to transmit a signal to two or more secondary transmitting stations. The secondary stations use this signal to synchronize their transmitters so that at a proper time interval each station transmits a signal. The difference in time of signal arrival is a measurement of the distance from a receiver to the ground stations. The points having the same measured difference in distance as a pair of stations form a hyperbolic line of position. Two or more of these lines intersecting define a position. This concept is shown in Figure 6.

Using a hand-held Loran instrument, the field person stands near the well

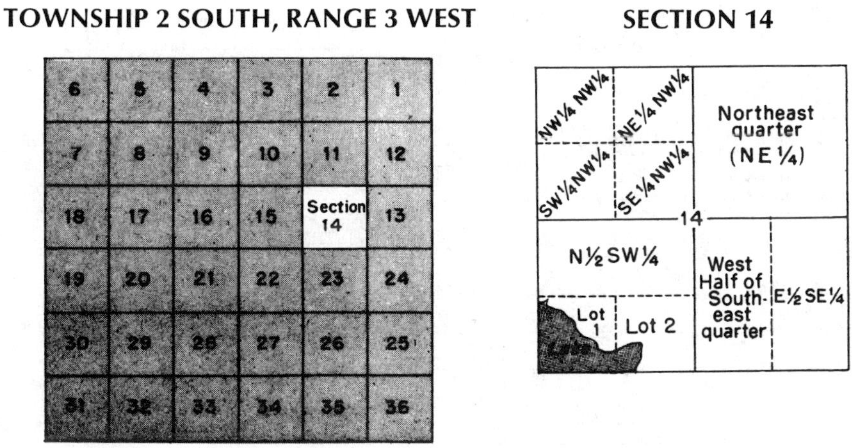

Figure 3.

location and records the latitude and longitude coordinates. These coordinates can then be converted[8,9] to the particular coordinate system (example: the State Plane system) used for the groundwater database. Most instruments available on the market should give an accurate reading within ±200 ft, which is suitable for databases that cover an area the size of a county. Figure 7 illustrates field use of the Loran-C hand-held instrument for obtaining the location coordinates for a newly installed private well. The 5-in. PVC well casing for a private water supply is also shown.

Other, more expensive field locational units are referred to as global positioning system (GPS) instruments. Instead of using signals from land-based stations, these instruments use navigational satellites. They offer a higher degree of accuracy than the Loran and also offer an elevation reading in addition to the latitude and longitude readings. This would provide an X, Y, and Z coordinate in one step. The major disadvantage of these units is the much higher cost over the Loran instruments. The cost would be prohibitive for most local agencies. For further information on these instruments contact the companies in Appendix D under "GPS."

Some items included on the well construction log can create problems for the mapper. A driller may document the builder or land developer as listed owner of the well, rather than the actual homeowner, and any attempt to locate such wells can be confusing and lead to many location errors. Well drillers do not always carry sectional maps into the field and may document the wrong section of a

GEOLOGICAL SURVEY SAMPLE No. 62061

7

WATER WELL RECORD
ACT 294 PA 1965

MICHIGAN DEPARTMENT OF PUBLIC HEALTH

1 LOCATION OF WELL

County	Twp.	Fraction	Section No.	Town	Range
Ingham	Vevay	SW ¼ SW ¼ NW ¼	21	2 N/S.	1 E/W.

Distance And Direction from Road Intersections
69' east of Hull Rd
36' north of Coy Rd.
Street address & City of Well Location

OWNER No.

3 OWNER OF WELL:
Address

2 FORMATION	THICKNESS OF STRATUM	DEPTH TO BOTTOM OF STRATUM
sandy clay	8	8
gravel ~~gray clay~~	7	15
gray clay	3	18
gravel	10	28
gray shale & sandstone	5	33
black shale	11	44
sandstone & shale strips	93	137
sandstone	18	155

4 WELL DEPTH: (completed) 155 ft. Date of Completion 6/25/69

5 ☐ Cable tool ☒ Rotary ☐ Driven ☐ Dug ☐ Hollow rod ☐ Jetted ☐ Bored ☐

6 USE: ☐ Domestic ☐ Public Supply ☐ Industry ☐ Irrigation ☐ Air Conditioning ☐ Commercial ☐ Test Well ☐ V.F.W. Post

7 CASING: Diam. Threaded ☒ Welded ☐ Height: Above/Below surface 1 ft.
4 in. to 46 ft. Depth Weight 11 lbs/ft.
in. to ft. Depth Drive Shoe? Yes ☒ No ☐

8 SCREEN:
Type: Dia.:
Slot/Gauze Length
Set between ft. and ft.
Fittings:

Figure 4. Top portion of a well construction record in Vevay Township, Ingham County, Michigan showing the section number, town, and range location, and the parcel fraction or quarter. Other information is also provided by the well driller such as owner, address, and location relative to the nearest roads. (From the Ingham County Health Department, Michigan.)

township, or even locate the well in the wrong county. It must be emphasized that local residents, such as local health department personnel, may provide the only solution to some of these location difficulties.

During the mapping process decisions must be made about the degree of accuracy. It is naïve to assume that all wells can be located within ±50 ft, especially when one considers the scale of the map. For many quadrangle maps, the dot drawn may already encompass 30 to 50 ft in diameter, depending on the scale of the map. To decide on an acceptable degree of error, database users must establish a list of uses for the database and ascertain what level of accuracy is needed for each. For uses such as creating geologic cross-sections or fence diagrams, ±200 ft may be acceptable.

Once wells are located on a topo map, the user must decide which geographical coordinate system for location will be assigned to the well construction records. The geographic graticule, or longitude and latitude system, is the most familiar. Longitude and latitude lines are based on two standard lines, the Equator at 0° latitude, and the Greenwich Meridian at 0° longitude. Lines of latitude and longitude are based on angular measurements of degrees, minutes, and seconds to

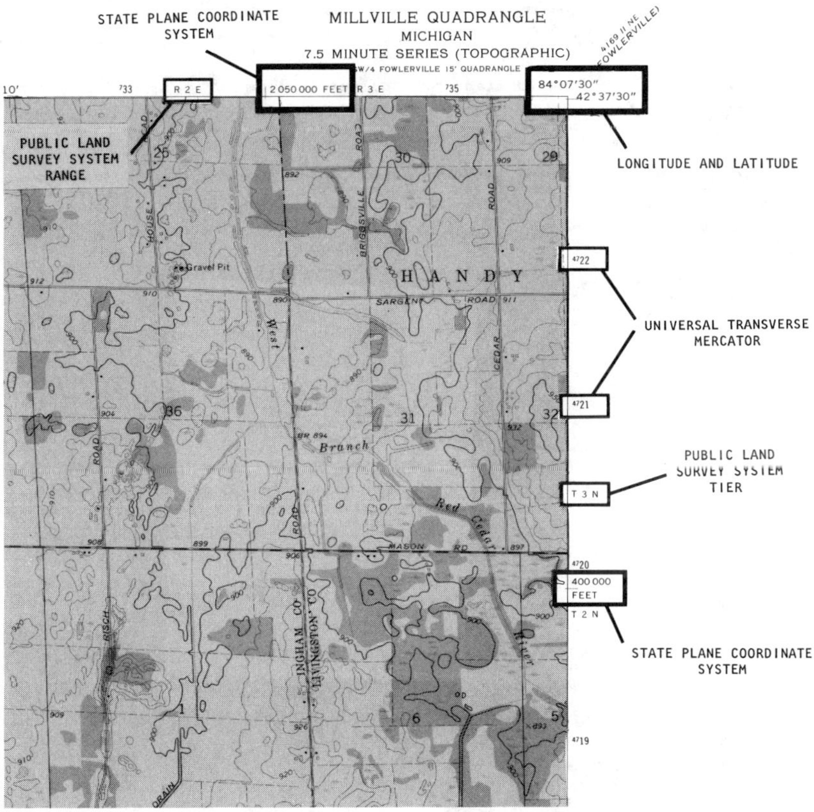

Figure 5. This upper right portion of a U.S. Geological Survey topographic map or Quadrangle indicates the location of three geographic coordinate systems and the tier and range survey system markings.

accommodate the spherical shape of the Earth. It is for this reason that ground distances are not constant from place to place. This system has the advantage of being used most often on a global scale and offers a common point of reference for many national and international groups and agencies. This system is also the one recommended by the 1987 EPA workshop data elements because other systems were not universal enough.

The Universal Transverse Mercator (UTM) system is a regular grid system, unlike longitude and latitude, because it is based on ground distances rather than angular measurements. While this system has the advantage of being a grid system its base unit is the meter, which is not commonly used in most land surveying and

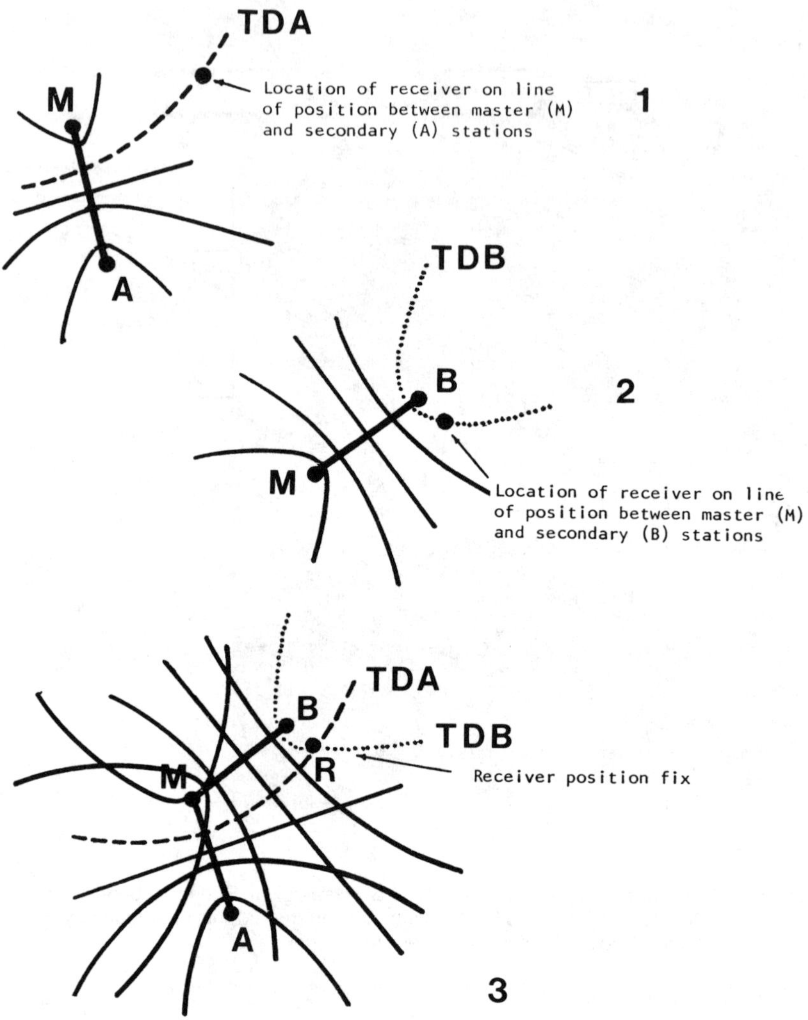

Figure 6. The concept of hyperbolic reception. The Loran-C instrument receives signals from a master and secondary stations to determine its geographic position. These readings are usually read as degrees, minutes, and 1/100 of minutes latitude and longitude. Software is available to convert these readings to State Plane or Universal Transverse Mercator coordinate systems.

Figure 7. Photograph of an Ingham County Health Department staff using a hand-held Loran-C instrument for obtaining latitude and longitude coordinates at the site of a private water well. The 5-in. PVC well casing is also shown in the foreground.

land descriptions in the U.S. In addition, the different zones that comprise the system are delimited by meridians of longitude which do not account for political boundaries such as county lines. This makes the system somewhat inconvenient for local database developers and users who must remain cognizant of their political neighbors.

For most database users, the State Plane Coordinate system is probably the most convenient because it is also a regular grid system based on ground distances rather than angular measurements. This system has the advantage over UTM in its use of feet as the unit of measurement and that different zone boundaries for the system take into account county boundary lines. An example of this mapping process uses the earlier example of the 1969 well shown in Figure 4. The well mapper could use the distance and direction information from the record in conjunction with a local health department permit for the on-site sewage system. Database users often have access to local health department records for both sewage and well permits (Figure 8). Site information from the well record and permit can also be matched and verified by a tax role or equalization table and a building permit. Using sewage and well permits, the location can be confirmed with either an aerial or plat map (Figure 9).

When reviewing well construction records along with other permit information during the mapping process, it is often useful to work in pairs (Figure 10). As one individual reviews the permit and well construction record in conjunction with aerial photos, the other person can match the data to the topographic map. Figure 11 demonstrates the use of aerial photographs to measure well location to various landmarks (road and section lines). This information is transferred to a topographic map (Figure 12). Locational data are then read from any one of the geographical location systems. For this example, the mapped well is shown in Figure 13 as well number 21011.

For other records, site or field verification may be necessary, especially where permit records, aerial maps, and plat maps are not available.

Regardless of the geographic location system used, most programs should be able to convert from system to system. One system can be reported using the computerized well construction record and a routine made available within the computer to complete the conversion automatically.

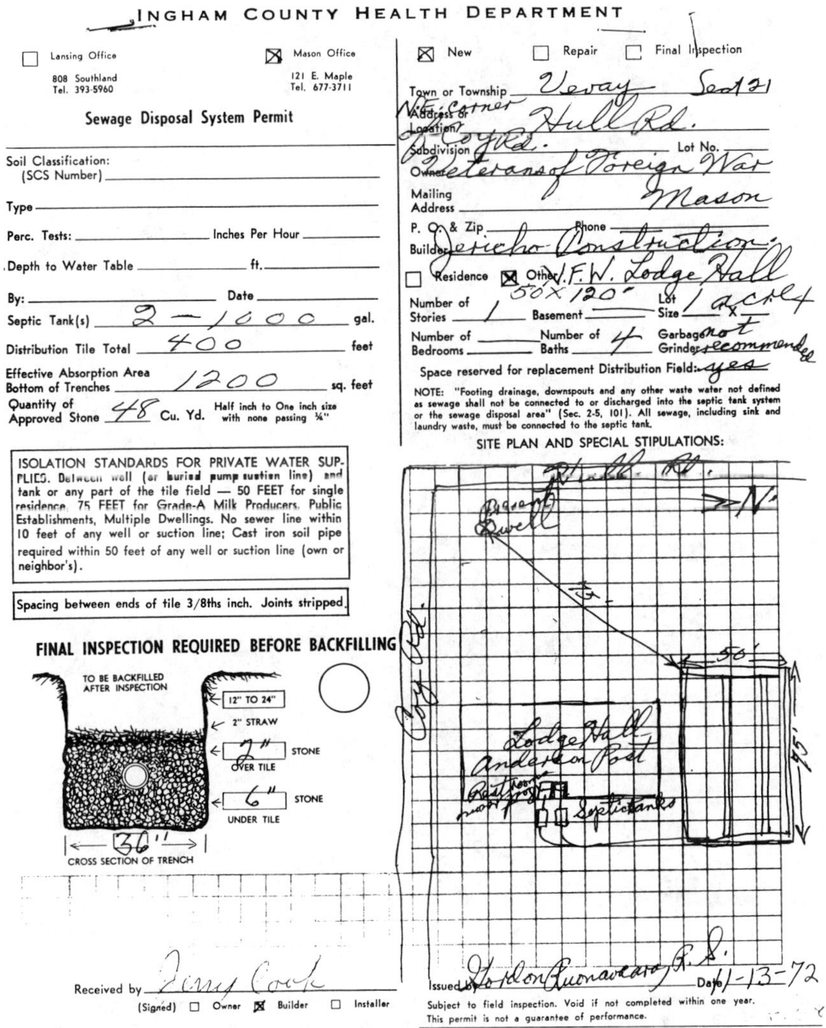

INGHAM COUNTY HEALTH DEPARTMENT

☐ Lansing Office
808 Southland
Tel. 393-5960

☒ Mason Office
121 E. Maple
Tel. 677-3711

Sewage Disposal System Permit

Soil Classification: (SCS Number)

Type

Perc. Tests: Inches Per Hour

Depth to Water Table ft.

By: Date

Septic Tank(s) 2 — 1000 gal.

Distribution Tile Total 400 feet

Effective Absorption Area Bottom of Trenches 1200 sq. feet

Quantity of Approved Stone 48 Cu. Yd. Half inch to One inch size with none passing ¼"

ISOLATION STANDARDS FOR PRIVATE WATER SUPPLIES. Between well (or buried pump suction line) and tank or any part of the tile field — 50 FEET for single residence, 75 FEET for Grade-A Milk Producers, Public Establishments, Multiple Dwellings. No sewer line within 10 feet of any well or suction line; Cast iron soil pipe required within 50 feet of any well or suction line (own or neighbor's).

Spacing between ends of tile 3/8ths inch. Joints stripped.

FINAL INSPECTION REQUIRED BEFORE BACKFILLING

TO BE BACKFILLED AFTER INSPECTION
12" TO 24"
2" STRAW
7" STONE OVER TILE
6" STONE UNDER TILE
36"
CROSS SECTION OF TRENCH

Received by Jerry Cook (Signed) ☐ Owner ☒ Builder ☐ Installer

☒ New ☐ Repair ☐ Final Inspection

Town or Township Vevay Sec. 21

Address or Location N.E. Corner Hull Rd. & Coyle Rd.

Subdivision Lot No.

Owner Veterans of Foreign War

Mailing Address Mason

P. O. & Zip Phone

Builder Jericho Construction

☐ Residence ☒ Other V.F.W. Lodge Hall 50 x 120'

Number of Stories 1 Basement Lot Size 1 acre +

Number of Bedrooms Number of Baths 4 Garbage Grinder not recommended

Space reserved for replacement Distribution Field: yes

NOTE: "Footing drainage, downspouts and any other waste water not defined as sewage shall not be connected to or discharged into the septic tank system or the sewage disposal area" (Sec. 2-5, 101). All sewage, including sink and laundry waste, must be connected to the septic tank.

SITE PLAN AND SPECIAL STIPULATIONS:

Issued by Gordon Guonavcaro R.S. Date 11-13-72

Subject to field inspection. Void if not completed within one year. This permit is not a guarantee of performance.

Figure 8. The well construction record in conjunction with the well and sewage permit issued by the local health department aids in the mapping process by providing preliminary township, section, fraction, road and site information. (From the Ingham County Health Department, Michigan.)

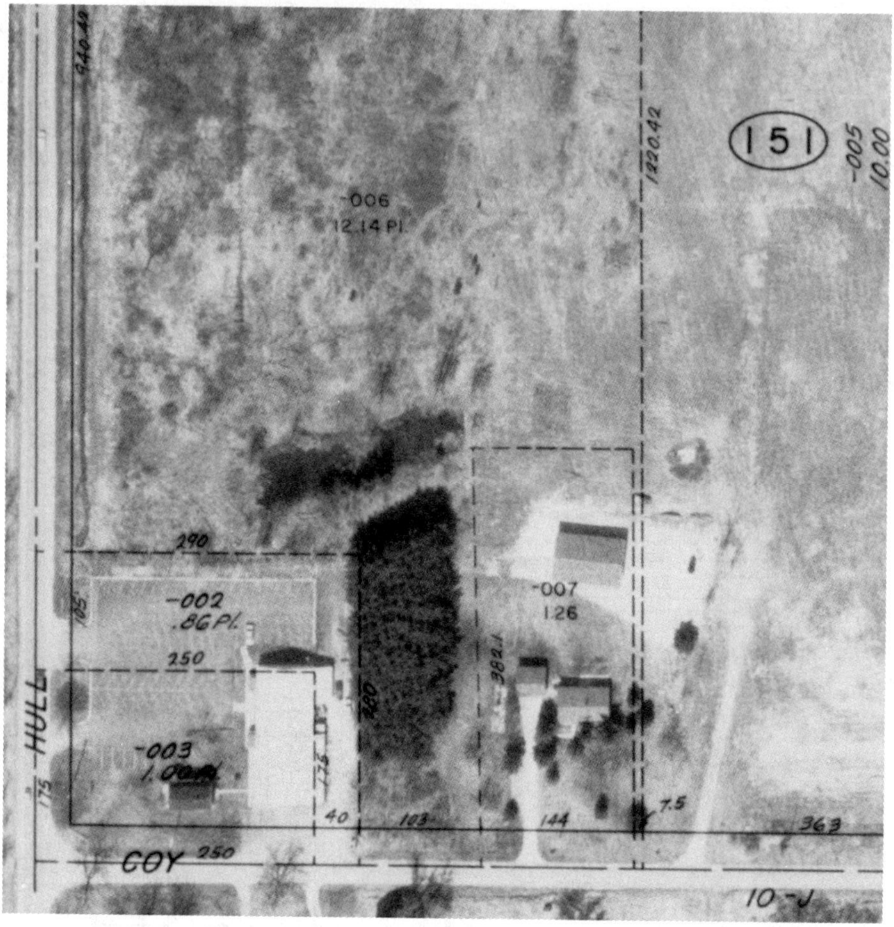

Figure 9. Aerial photo of site for well location demonstrated in both Figures 4 and 8. These aerial photos are used to help confirm the location of the well in respect to roads, buildings, and other landmarks. From this aerial, the approximate well location indicated by the health department permit see (Figure 8) is measured and the information is then used to place the location on the U.S. Geological Survey topographic map (see Figure 10).

Figure 10. Information from the permits used to place the probable location of the well on the aerial photographs.

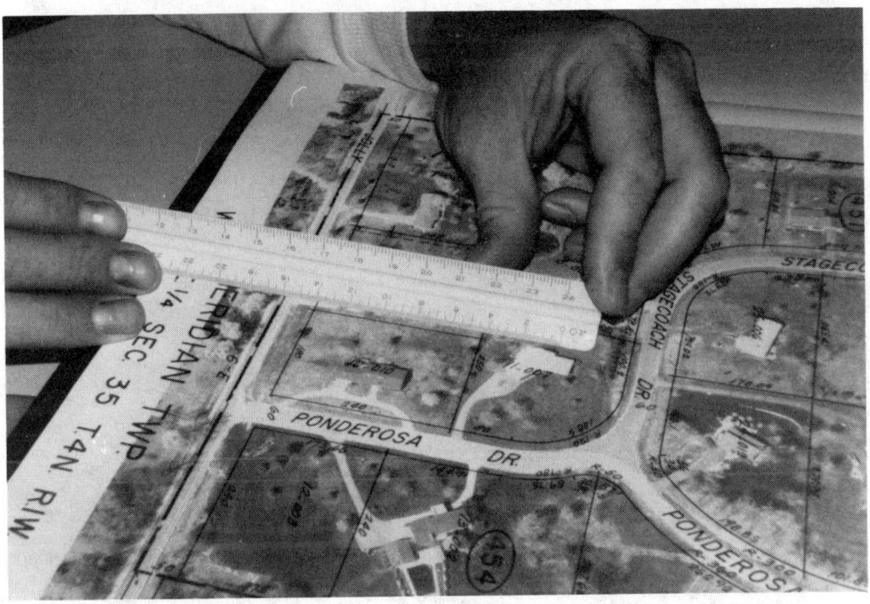

Figure 11. Using the aerial photograph to measure the location of the well relative to landmarks such as roads.

Figure 12. Placing the well location on a topographic map in preparation for digitizing.

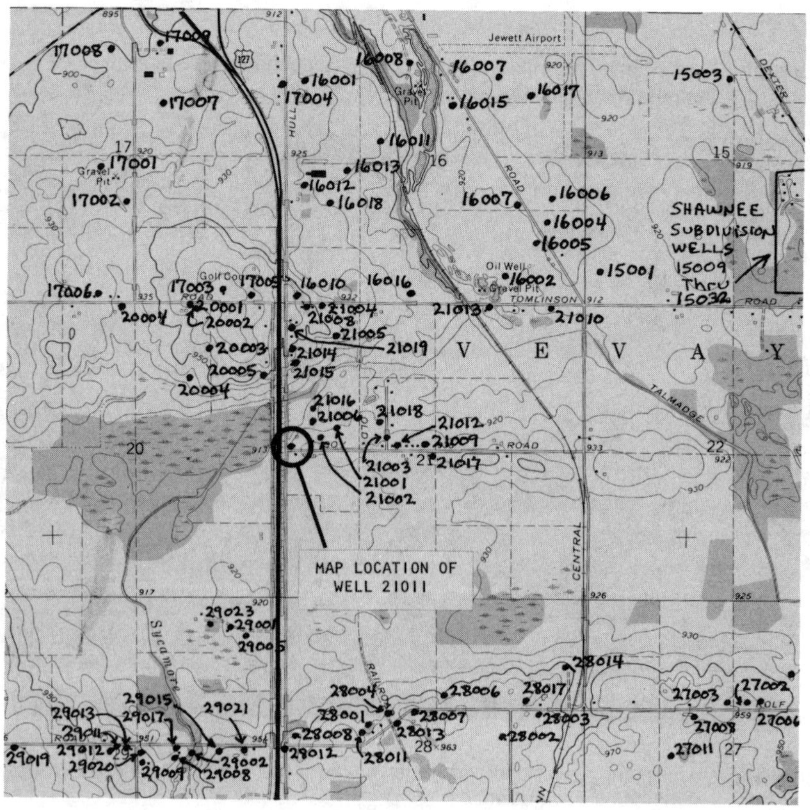

Figure 13. A portion of the U.S. Geological Survey topographic map showing the marked location of well 21011, or the well demonstrated from Figures 10 and 12.

REFERENCES

1. Forstat, D. W., and D. A. Walsh. Geological Survey Division Groundwater Database Pilot Project. Michigan Department of Natural Resources, Geological Survey Division (October 1986), Table 1.
2. Thompson, G., D. Walsh, and M. Beaulac. *Water Well Record Location Verification Methods: A User Manual,* Michigan Department of Natural Resources, Land and Water Management Division (1989).
3. Beaulac, M. N. "Statewide Groundwater Database Strategy: What is Identified and What is Being Done About It," *Policy, Planning and Resource Protection - A Groundwater Conference for the Midwest: Proceedings* (Kalamazoo: Western Michigan University, 1988), p. 86.
4. Monaghan, G. W., and J. L. Grahame. "Groundwater Computer Notes: A Computerized Groundwater Resources Information System," *J. Assoc. Groundwater Sci. Eng.* 23(2) (1985).
5. Statewide Groundwater Data Base: Program Perspectives and Summary and Statewide Groundwater Data Base Strategy. Michigan Department of Natural Resources, Land and Water Management Division (1989).
6. Lusch, D. P. "The Fundamentals of GIS Workshop, Center for Remote Sensing and Dept. of Geography, Michigan State University, East Lansing, MI 1988. Appendix H," in *Restoration of Lost or Obliterated Corners and Subdivision of Sections: A Guide for Surveyors,* U.S. Department of the Interior, Bureau of Land Management, Division of Cadastral Survey (1974).
7. Dull, C. W. Loran-C Radio Navigation Systems as an Aid to Southern Pine Beetle Surveys, Agric. Handb. No. 567, U. S. Department of Agriculture Combined Forest Pest Research and Development Program (1980).
8. Core Con — File or Point Conversion Software for Michigan to Convert Lat-Long Coordinates to State Plane, Michigan Department of Natural Resources (1989).
9. Plane-PC Interactive Grid Coordinate Coversion for Personal Computers, National Mapping Program Technical Instructions, Department of the Interior, U.S.G.S. National Mapping Division (1984).

CHAPTER 4

Tools – Hardware, Software, Experience

Building a computerized groundwater information system presents new challenges for the environmental professional. Skills must be learned, and new budget line items must be justified. The importance of finding experienced people cannot be overstated. Based on the experiences of others, the following general guidelines will help prepare any organization and its budget.

1. RULES OF THUMB

General rules apply to database projects and related geographic information systems (GIS). Although these are often dismissed before the project is begun, they are widely acknowledged as true afterward.

- Make a 3-year implementation plan and save half of the hardware/software money for the second year.
- Outlays for a successful project:
  15 to 20% hardware and software
  80 to 85% creating the database
- Make the support/training budget twice the hardware/software budget. Plan support/training to take twice as long as the actual installation time. Find experienced users to help, and budget for staff training.
- It usually takes 30 to 40 hr to become proficient with a particular software package. In the workplace, that 40 hr may extend over 2 months. With such a long period of time involved, one should consider buying software in phases. This will save money as well as an updated version by waiting to buy it until the staff has learned the original package.

- A realistic project schedule recognizes these five implementation stages:[1]

  1. Original system acquired, great expectations (12 to 18 months)
  2. Realization, anxiety, depression (1 to 3 months)
  3. Learning what can be done (12 to 18 months)
  4. Specific projects, pilot projects and evaluation (12 to 24 months)
  5. System implementation

2. A MINIMAL SYSTEM

On a very basic level, electronic files can be created using a PC and a spreadsheet. For example,

TOWNSHIP: Thornapple SECTION: 06

Address	Depth	Aquifer	Static Water Level	Elevation
12225 Bass Rd.	74	Thornapple	60	830
1394 Patterson	86	Thornapple	50	850
1395 Patterson	71	Thornapple	50	850
RR 1 Wayland	75	Thornapple	51	855
1550 Patterson	64	Thornapple	38	830

The total hardware/software investment may be PC, two floppy drives, monitor, printer, spreadsheet = $1100; however, for all practical purposes, this system is meager, although it is preferable to no system. A modest and expandable system may include: AT-286 compatible computer, math coprocessor, 40 Mb hard disk, database manager, word processor, mapping/graphics software, printer = $6000.

3. AN IDEAL SYSTEM

When an organization is deeply involved in groundwater management, its information base is much more detailed. Digital files for various attributes may be stored in the database — lakes, rivers, streams, highways, land use, contamination sites, and wells. Each digital file can add a layer to the maps and a variable to the analyses. Any organization with an eye to growth and better management will wish to purchase more hard disk storage, a printer, and a plotter. A good system for a groundwater management specialist may include such hardware as 386- or 486-based computer, math coprocessor, 320 Mb hard disk, 1.2 Mb floppy disk, 640K RAM + 8 Mb expanded memory, VGA graphics card, monographics card, color monitor, monochrome monitor, two serial ports, two parallel ports, laser

Figure 1. Basic components of a groundwater information system. Clockwise from top: computer, digitizer, plotter, printer.

printer, plotter, 24 × 36 digitizer, mouse, and modem. The software packages may include digitizing, word processor, database manager, spreadsheet, graphics manager, map/display, cross sections, GIS, statistics and communications. The total cost for such a system is $15,000 to $30,000. Figure 1 shows a generic workstation for a groundwater database or GIS.Appendix D lists software tools used for examples in this book.

REFERENCES

1. Brown, C. *Implementing a G.I.S. - What Makes a New Site a Success,* Environmental Systems Research Institute, 380 New York St., Redlands, CA 92373.

CHAPTER 5

Data Entry – Electronic Files

When data are ready to be entered into the computer, many decisions will have been made about the information to be collected. The level of preciseness of the database, what codes will be used, and an idea of the types of errors (keyboard entry and interpretation errors) that may require attention will have been considered; however, ideas will, no doubt, be revised after five or six well records have been entered.

1. GENERAL ATTRIBUTES OF ELECTRONIC DATABASES

To build electronic files, use a database software package or database manager. Some common commercially available database managers are dBase,[1] Foxbase,[2] Rbase,[3] PCFile,[4] Vulcan,[5] Quattro,[6] Q & A,[7] etc. Database managers offer several advantages:

- Data are accessible to nonprogrammers: it is easy to learn how to enter and retrieve data.
- File structures can be easily altered: a nonprogrammer can design a file, and can modify its structure. Most files structures are changed before the final version is agreed upon.
- Universal use means many enhancements are available: because database software is used so extensively, many companion products can be found that add ease or features. For example, data can be easily imported or exported to other software packages.

2. SELECTING THE BEST DATABASE MANAGER

The selection of a database manager is a matter of taste. Many have similar features, but prices can vary greatly. For example, database managers using the dBase programming language and file structure range in price from $99 to $1200.

Select database software after considering:

- Can good support be obtained?
- Is it familiar to the team?
- Is import and export to other software convenient?
- Is error checking easy to incorporate?
- Is it used in other jurisdictions for the same purpose?

A small number of database managers, software, and user manuals exist for groundwater. In Michigan, a dBase-style file is available through the Department of Natural Resources.[8] By using their files, software, and manuals, weeks of programming and documentation time can be saved.

3. DESIGNING A DATABASE

If a database is designed from scratch, one must balance the desire to include as much information as possible with the practicalities of project budget and timetable. One must also consider other potential users of the database so that something they may find invaluable is not omitted. Sometimes a compromise can be reached by including all the fields in the file structure, but inputting the data on a priority basis. For example, the entire content of a well-log can be contained in the file structure, but local groundwater sites may enter the most useful information first (location coordinates, well depth, stratigraphy), and add the bulk of the well-log information to the records as time allows (pump manufacturer's name, driller permit number, etc.).

Although it is possible to revise a database's structure, it is best to start with a well-planned structure. If a file is constantly revised, more trouble will occur at entry time because less time is available to check for keyboarding errors, and the system may be plagued with converting data from one file structure to the next.

The larger the planned database, the more attention should be directed to eliminating keyboard and interpretation errors. It is a good idea for the manager to type/enter in a few records, andto solicit feedback from those who input the records. Quality control is very important during input. One can create and run error check programs that look for inconsistencies and common errors, but this is best used as a final check on good keyboard entry procedures.

4. SPECIAL FEATURES OF A GROUNDWATER DATABASE

The following four features are particularly important to groundwater concerns:

1. Quality of the original well-log itself can be subjective. It may be beneficial to create a database field for confidence; A means a top-quality, professionally done engineering wel-log, B means a recently inspected new well with well-log supplied by a competent well driller, C is average, and D means questionable.
2. Geologic interpretation of lithology may vary from driller to driller. Procedures are needed to code each lithology and to double-check this code against the description of the well driller. In the Michigan database, a lithology code system consists of 60 general code categories to classify all possible well driller descriptions.
3. The details of well depth and screen placement have some subtleties that may need to be captured. For example, a driller may drill to a depth of 150 ft, yet place the screen at 100 ft. How deep is the well? To study lithology, information to 150 ft is useful, but to estimate where groundwater came from, it is best to consider where the screen is placed, not only how deep the hole was dug.
4. A nonduplicated well ID number should be assigned so that related files can be linked efficiently. The address or well owner name can change, therefore it is better to assign a specific ID number for each well. This well number allows you to link well construction, lithology, and chemistry data about the well.

5. AN EXAMPLE OF A WELL CONSTRUCTION DATABASE

Michigan's database structure evolved over six revisions. The current version has been used to input, report, and display tens of thousands of drillers' well logs. Wellkey software is written and compiled in Foxbase, a dBase-style database manager. By compiling the Wellkey programs, the State of Michigan can distribute them to users who may not have database managers of their own. Figure 1 shows the initial screen for Wellkey. Table 1 shows the PC-based file structures for Michigan's Wellkey (well construction database).

6. THE TEAM APPROACH TO DATA ENTRY

If an organization expects its database to be large, special attention should be paid to team assignments during data entry. Based on Michigan's collective experience of computerizing 37,000 well records in 6 years, the following is a good order of events:[8,9]

```
Input File:DEMO                    Record:     1 of 3        Updated:05/24/89

1 LOCATION OF WELL   <<<33040101001>>>

County           Township Name       Fraction              Section  Town    Range
 INGHAM           MERIDIAN            SWQQQS NWQQS NWQS     01      04 N    01 W

Distance & Direction From Road Intersections   3 OWNER OF WELL
  0.45 MILES E OF SHOESMITH RD, 150' E OF GR    Name SPALDING, GORDON
  6188 GREEN RD.      HASLETT   48840-          Add1 6188 GREEN RD.   2
Street Address & City of Well Location          City HASLETT   MI 48840-
                                                Address Same as Well?  Y
Loc E 1964318.25  Coord Sys S  Verified F
Loc N  454680.12        Mtd D  Source   S       Tax Parcel No.

2 FORMATION DESCRIPTION Thkness Depth  Code  4 WELL DEPTH Date Compl New or
  1 CLAY                   10      10    10                Mo Da Yr  Replce
  2 SAND                   10      20    20     235.0 ft   12 11 69    U
  3 GRAVEL                 10      30    30
  4 LOAM                   10      40    40  5 RIG  H      6 USE  DOM
  5 SANDSTONE              95     135    50
  6 SHALE                 100     235    55  Elev  880.00 Mtd T Dry Hole?N
```

Figure 1. Basic entry screen for Wellkey, a database program for the entry of Michigan well records. (From Wellkey, Statewide Groundwater Database, Michigan Department of Natural Resources, Land and Water Management Division, 1989.)

Table 1. Michigan's Wellkey File Structure(s)

Structure for database: E:\WELLKEY\DEMO.LAB
Number of data records: 3
Date of last update: 05/24/89

Field	Field Name	Type	Width	Dec
1	WELLID	Character	11	
2	CNTY	Character	2	
3	TWP	Character	4	
4	QQQS	Character	2	
5	QQS	Character	2	
6	QS	Character	2	
7	SEC	Character	2	
8	TIER	Character	2	
9	N_S	Character	1	
10	RANGE	Character	2	
11	E_W	Character	1	
12	LOC_E	Numeric	10	2
13	LOC_N	Numeric	10	2
14	COORD_SYS	Character	1	
15	COORD_MTD	Character	1	
16	VERIFIED	Character	1	
17	DIRECTIONS	Character	55	
18	ELEV	Numeric	7	2
19	ELEV_MTD	Character	1	
20	WELL_ADD	Character	30	

Table 1 (Continued). Michigan's Wellkey File Structure(s)

Field	Field Name	Type	Width	Dec
21	WELL_CITY	Character	17	
22	WELL_ZIP	Character	5	
23	WELL_ZIP2	Character	4	
24	OWNER_NAME	Character	30	
25	OWNER_ADD1	Character	30	
26	OWNER_ADD2	Character	30	
27	OWNER_CITY	Character	17	
28	OWNER_STAT	Character	2	
29	OWNER_ZIP	Character	5	
30	OWNER_ZIP2	Character	4	
31	ADD_SAME	Character	1	
32	TAX_NO	Character	12	
33	DEPTH	Numeric	5	1
34	MONTH	Character	2	
35	DAY	Character	2	
36	YEAR	Character	2	
37	NEW_WELL	Character	1	
38	RIG	Character	1	
39	USE	Character	3	
40	DRY_HOLE	Character	1	
41	C_TYPE	Character	1	
42	C_JOIN	Character	1	
43	C_DIAM1	Numeric	5	2
44	C_DEPTH1	Numeric	5	1
45	C_DIAM2	Numeric	5	2
46	C_DEPTH2	Numeric	5	1
47	H_DIAM1	Numeric	5	2
48	H_DEPTH1	Numeric	5	1
49	H_DIAM2	Numeric	5	2
50	H_DEPTH2	Numeric	5	1
51	C_HGTH	Numeric	4	1
52	D_SHOE	Character	1	
53	S_TYPE	Character	1	
54	S_DIAM	Numeric	5	2
55	SLOT_GAUZE	Character	1	
56	S_OPENING	Numeric	6	3
57	S_LENGTH	Numeric	4	1
58	S_DEPTH1	Numeric	6	2
59	S_DEPTH2	Numeric	6	2
60	FITTINGS	Character	2	
61	S_ABOVE	Numeric	3	1
62	SWL	Numeric	6	2
63	FLOW	Character	1	
64	PUMP_LEV1	Numeric	4	
65	PUMP_HRS1	Numeric	4	1
66	PUMP_RATE1	Numeric	4	
67	PUMP_LEV2	Numeric	4	
68	PUMP_HRS2	Numeric	4	1
69	PUMP_RATE2	Numeric	4	
70	WELL_HEAD	Character	1	
71	GROUT_TYPE	Character	1	
72	CNTM_TYPE	Character	2	

Table 1 (Continued). Michigan's Wellkey File Structure(s)

Field	Field Name	Type	Width	Dec
73	CNTM_DIST	Numeric	4	
74	CNTM_DIREC	Character	3	
75	DISINFECT	Character	1	
76	PLUGGED	Character	1	
77	PUMP_IN	Character	1	
78	PUMP_TYPE	Character	1	
79	PUMP_MAKER	Character	20	
80	PUMP_HP	Character	4	
81	PUMP_PIPE	Numeric	3	
82	PUMP_GPM	Numeric	4	
83	DRILLER_NO	Character	5	
84	OPERATOR	Character	30	
85	DATA_SOURC	Character	1	
86	ENTRY_MO	Character	2	
87	ENTRY_YR	Character	2	
88	CHARHEIGHT	Numeric	5	1
89	CHARWIDTH	Numeric	5	1
90	CHARANGLE	Numeric	5	1
** Total **		535		

Structure for database: E:\WELLKEY\DEMO.LTH
Number of data records: 20
Date of last update: 05/24/89

Field	Field Name	Type	Width	Dec
1	WELLID	Character	11	
2	FORM	Character	72	
3	THK	Numeric	3	
4	BOT	Numeric	4	
5	SC	Numeric	2	
** Total **			93	

Structure for database: E:\WELLKEY\DEMO.REM
Number of data records: 1
Date of last update: 05/24/89

Field	Field Name	Type	Width	Dec
1	WELLID	Character	11	
2	COMMENTS	Character	254	
** Total **		266		

From Wellkey, Statewide Groundwater Database, Michigan Department of Natural Resources, Land and Water Management Division (1989).

1. Screen the paper records, use those that have the essential information (well depth, static water level, location information, lithology descriptions). If interest lies in the geological aspects, one may select records for deep wells because they contain more geological information. If interest lies in wells that may be affected by road salt application, more shallow wells should be included. Ask a local expert for advice in selecting the most useful wells for the project.
2. Organize good records into file folders by township and sequence each township file by section number. In general, townships should be treated as the team-unit; i.e., if the staff is limited, complete one township at a time. If the project is staffed by a good number of people, team them according to townships.
3. Dots should be placed on a topographic map for each confirmed well location. The well number should be written on the topographical map next to the dot and on the paper log. An accurate well location dot on a topographic map is essential, and the help of local experts is critical.
4. Digitize all the well dots to create location coordinates. For quality control, ask one person do all the digitizing because setting up the map on the digitizer, developing accurate entry techniques, etc. are quite detailed. This will help build consistency across townships' data. Figure 2 shows this process. Elevation data are also obtained from the topographic map, so it is convenient to enter elevations into the database. C-Map, the particular well digitizing package used in Michigan, has options to key in elevation during location digitization.
5. Accurate location and digitizaton are essential to the project, so the team leader should have final approval and compare a map plotted from the ditigized data with the original topographic map with dots. Any improperly placed points should be corrected. U.S. Geological Survey topographic maps of 7.5 min quadrangles are accurate to about ±40 ft. If more precision is desired, plat maps or aerial photographs which have been georeferenced to topographic map coordinates should be used.
6. Enter the rest of the information on the well to the database. This job should be given to fast, accurate typists who work on one township at a time. If the software used has good error checks, help screens and function-key shortcuts, the keyboard entry phase can be accurate and efficient.

The paper logs may need to be marked with any codes chosen for use. For example, a geologist could place lithology codes next to lithology descriptions so the keyboard operator could input them without having to pause to interpret. In the Michigan system, the keyboard operator enters the code and types the driller's actual description. A checker program can then search for inconsistencies between the code and the description. When the electronic record is created, print the wellid number, location coordinates, and elevation on labels and attach them to each matching paper log.

Figure 2. Process of digitizing well locations from topographical map to computer database.

REFERENCES

1. dBase III Plus and dBase IV are a registered trademark of Ashton-Tate.
2. Foxbase is a registered trademark of Fox Software, Inc.
3. Rbase is the registered trademark of Microrim.
4. PCFile is the registered trademark of Buttonware.
5. Vulcan is the registered trademark of Ratliff Software Production, Inc.
6. Quattro is the registered trademark of Bordland International.
7. Q & A is the registered trademark of Symantec.
8. Morrison, B., and C. Grobbel. *Wellkey User Documentation*, Michigan Department of Natural Resources, Land and Water Management Division (1989), p. 46.
9. "Well-Log Database Project: Data Collection and Quality Assurance Procedures," Grand Valley State University Water Resources Institute, Allendale, MI (1989).

CHAPTER 6

Geology of the Database Unit – What's Down There

Once a database of well construction records or well-logs is created, the data can be used to illustrate surface and subsurface features and to describe the flow characteristics of groundwater. The first step to understanding the hydrogeology of an area is having an idea of the groundwater flow patterns and the material water must travel through. This is part of the process of groundwater management and protection.

1. THE CROSS-SECTION

Just asan X-ray can be used to describe the interior structure of the human body, subsurface diagrams of lithology (cross-sections) can be used to describe the subsurface structure of the land. These cross-sections are created from the lithology data provided by the well-log records in the database.

In the past, cross-sections and fence diagrams were drawn by hand. By looking at several well-logs, a draftsman would draft or draw on paper the manner in which the different lithology layers were structured and arranged in relation to one another. Drawing these diagrams was very time consuming and tedious. For any given project, a limited number of cross-sections would be created by hand because of time, money, and personnel restraints. As a result, an inadequate number of cross-sections may be used to describe the hydrogeology of an area.

With appropriate software tied into the well-log database, an adequate number of cross-sections can now be created faster and more precisely than in the past.

Several software programs are available that provide the hydrogeologist with the appropriate graphics. The software program for illustrating cross-sections in this text is the lithology database and graphics program developed by Hall Ground-Water Consultants located in St. Albert, Alberta, Canada.[1]

2. AN EXAMPLE

Figure 1 represents a portion of the well location map for Vevay Township in Ingham County, Michigan. Well locations are indicated for those well records chosen for a cross-section diagram. All the well-log records for all the wells noted on this map are also contained in a larger well-log database for Ingham County.[2] To describe this area by use of cross-sections, several lines of connected wells must be identified, usually in a north-south, east-west grid pattern. The number of cross-sections made for any given area depends on the density of located wells on the map and the identified goals of the particular user.

Once an appropriate number of cross-section lines have been identified on the map, the computerized database can retrieve the necessary well records for each cross-section. All the information needed for the cross-section (location coordinates, well head elevation, lithology, static water level, and well I.D. number), are exported from the well-log database to the cross-section software. The area identified for cross-section, Figure 1, shows a line of wells approximately 3 3/4 mi long, including seven wells. Figure 2 represents the cross-section from Figure 1, drawn with the Hall Geobase4 software.[1]

Each study requires a different scale appropriate to the identified goals of the user. The cross-sections shown in the Vevay Township example may not be suitable for a smaller study which could only include a section of the township. The greater the need for detail, the smaller the scale required for observing more specific changes in geologic information. A series of cross-sections can be created for Veray Township and in conjunction with topographic maps, give a general geologic description for selected areas of the township. Figure 3 shows one cross-section with notes by the interpreter or geologist, indicating areas of hydrogeologic importance.

Once the series of cross-sections are reviewed, a picture of the lithology for Vevay Township can be created showing how the aquifers used for well water are arranged. The places at which aquifers may be open to surface water infiltration or groundwater recharge can also be geographically located, giving the user a map of sensitive geology. This map can be used to point out areas in the township at which the groundwater is vulnerable to pollution by chemical spills or possible sites of environmental degradation such as a landfill.

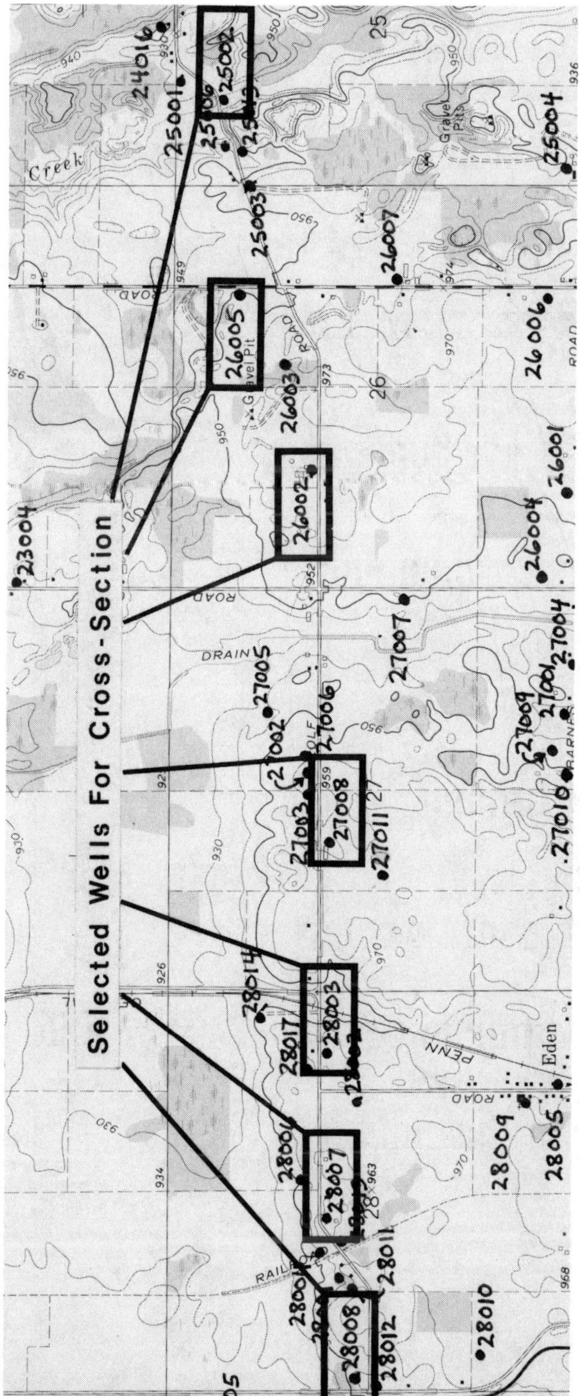

Figure 1. A segment of the prepared well location topographic map in Vevay Township, Ingham County, Michigan. A group of 7 wells was chosen to create a west to east cross-section. The well information is taken from the well-log record database to provide the lithology data.

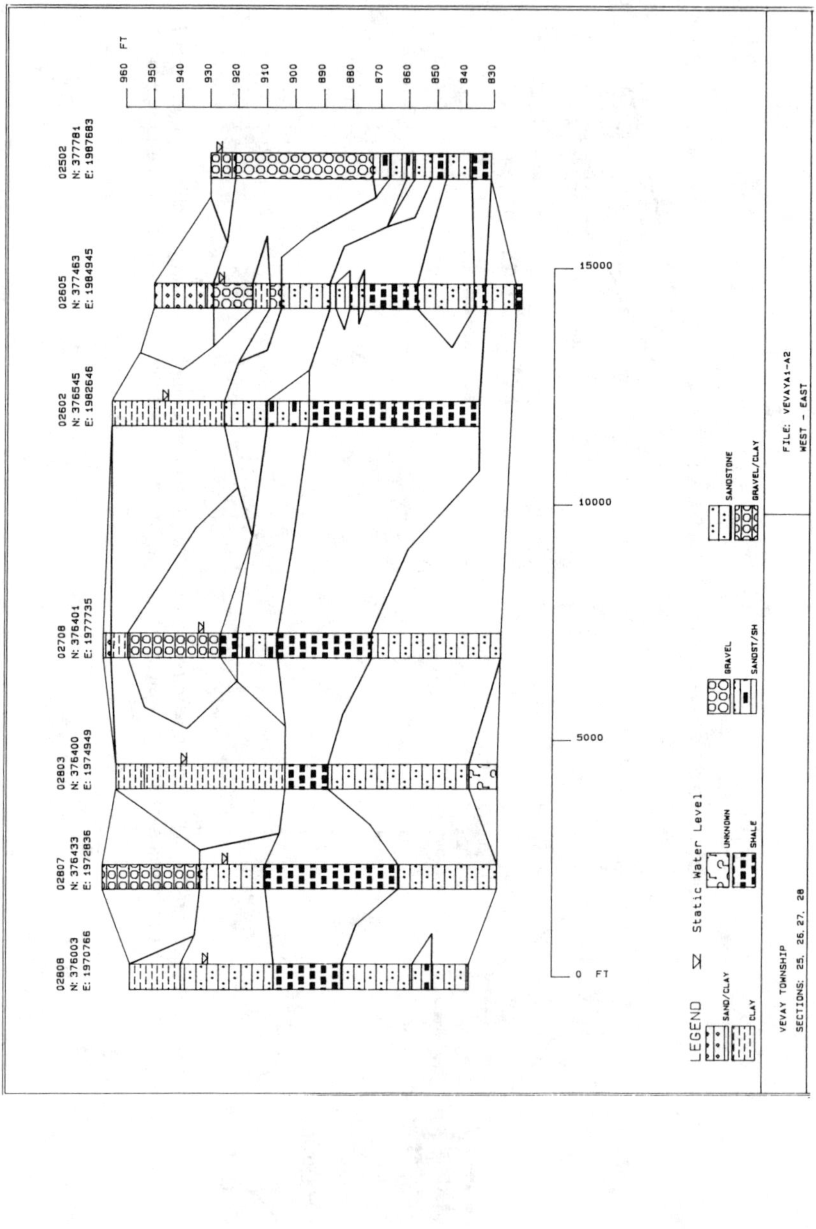

Figure 2. Cross-section of the wells chosen in Figure 1 created from data out of the well-log database and drawn from the Hall Groundwater software (Geobase4). Common lithology types are connected together by lines placed into the program by the database user.

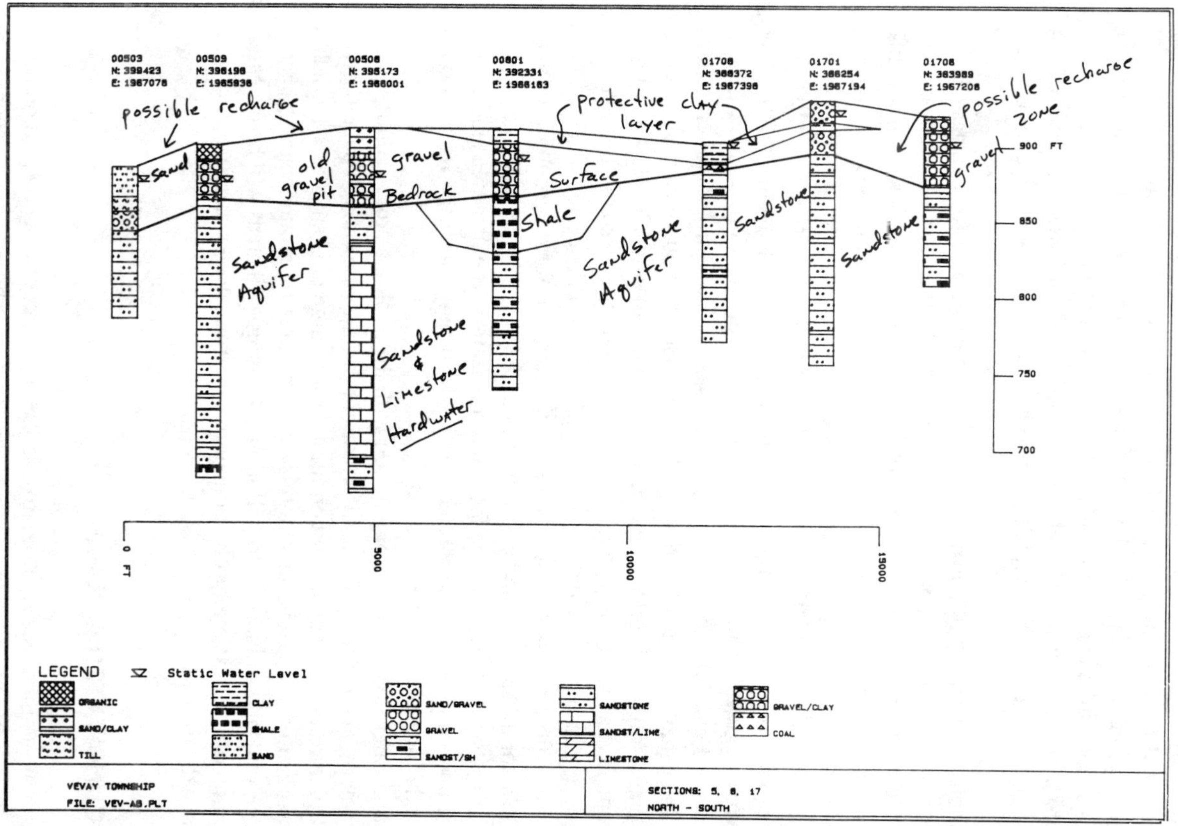

Figure 3. Another cross-section from Vevay Township that includes some notes of site interest from the database user. Cross-sections used in conjunction with other information help the database user create a picture of an area's general hydrogeology.

3. THE STATIC WATER LEVEL CONTOUR MAP

Groundwater flow direction is usually determined by measuring the static water levels in wells located in the same aquifer that have become equilibrated with the aquifer after construction. By plotting these static water levels on a map and connecting points of equal hydraulic head, a static water level contour map can be created to show groundwater flow direction for a given area.

Using a well construction database, the information necessary to create a static water level contour map can be transferred to a software package that will produce the appropriate map in a relatively short time. The following example uses a software program called Surfer,[3] developed by Golden Software, Inc. in Golden, CO.

In employing the Vevay Township example again, the well-log database can be called on to provide static water level data, well head elevation (approximate ground elevation in feet above sea level), location coordinates, and well identification numbers. Static water level data are subtracted from the well head elevations to give the static water level data in feet above sea level. The location coordinates (X and Y) and the static water level (Z) data are transferred to Surfer software. The software is then used to generate a static water level contour map as shown in Figure 4. This map may be used by a hydrogeologist to show general groundwater flow direction, water table depressions, etc., for an aquifer system.

It is important to remember that the final product or map created by both the database and software is only as accurate as the data collected. Static water levels measurements can fluctuate severely over time and by the season of the year. One may choose to use only the data collected for a certain time period and season. A well construction database provides a suitable database for determining groundwater flow for an area the size of a township, but the data resolution may not be suitable for a smaller area such as the township section. Smaller area studies may require more compact or dense data points which may mean the installation of more wells for more accurate data.

Groundwater flow maps in conjunction with the geologic cross-sections, topographic maps, and other available information, can provide enough data to give a general picture of hydrogeologic conditions for a large area such as a township. This information can assist in pointing out locations of groundwater recharge, discharge, and sites vulnerable to pollution. Mapping such areas will be necessary for creation of the next major component of the groundwater database, the water chemistry database.

4. GROUNDWATER MODELS

Groundwater models were used to describe or predict hydrogeologic conditions before microcomputers. Some models started in a physical form, such as an aquarium filled with sand. Many useful models that involve calculations have

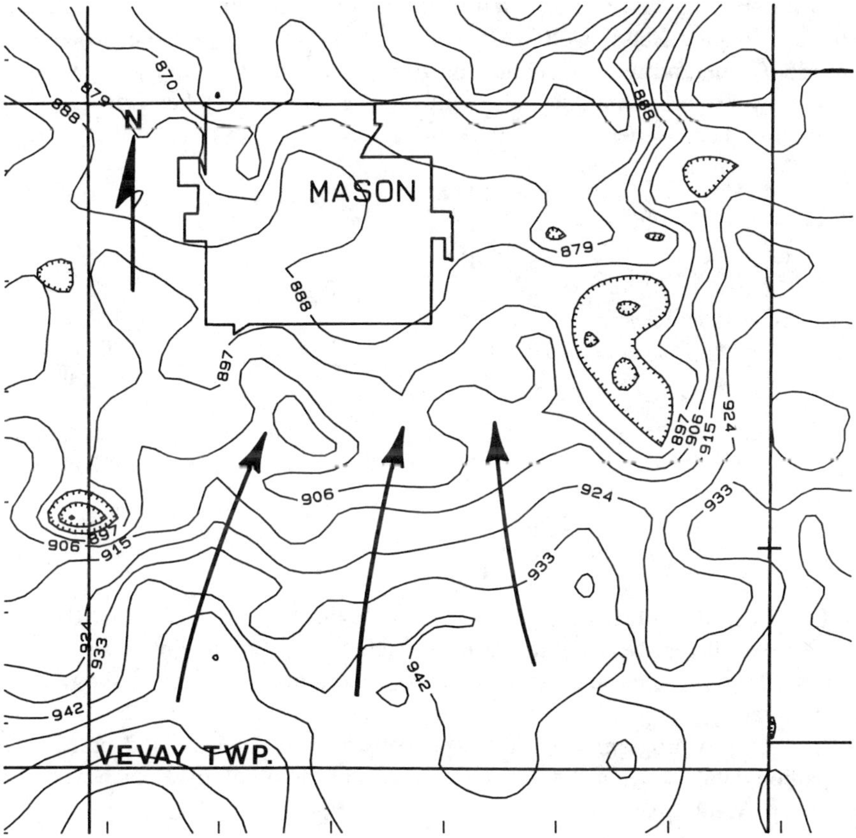

Figure 4. A static water level contour map created by using the data from the well-log database and the Surfer program. The general direction of groundwater flow shown by the arrows is to the north-northeast. Because the data are from bedrock wells this groundwater flow direction represents only the bedrock aquifer and not the more shallow glacial drift aquifers on top.

been converted to microcomputer programs. Some common microcomputer models are groundwater flow, groundwater vulnerability, and solute transport. Many of these models require actual field data along with some assumptions to help complete a picture of the hydrogeology.

5. AN EXAMPLE MODEL: AQUIPRO

Models can be devised to match particular databases. For example, Dr. Richard Passero of Western Michigan University's Geology Department developed Aquipro to use information available in the Michigan Groundwater Survey[4] data files. Aquipro analyzed and reported well depth and lithology details, applied a formula, and produced a well analysis file of information pertinent to aquifer protection/vulnerability. The higher the Aquipro score, the more protected the well. Below are three example wells with varying lithologies taken from the well construction database:

	Well #1	**Well #2**	**Well #3**
Well ID#	080110101	080110102	080010103
Depth	100	100	112
Screen Placement			95-100
Lithology	20 Sand	100 Clay	30 Sand
	35 Clay		60 Sand/gravel
	50 Sand/clay		100 Gravel
	55 Sand		112 Clay
	60 Gravel/clay		
	70 Clay		
	100 Sandstone		

The first well (080110101) is a mixed lithology, the second (080110102) is a theoretical well representing an extreme in clay-layer protection, and the third (080010103) represents an extreme in unprotected lithology. Figure 5 shows each of these wells.

The Aquipro program reads the well construction database and prepares a companion database which contains an analysis of the lithology of each well and assigns an Aquipro score.

	Well #1	**Well #2**	**Well #3**
Well ID #	080110101	080110102	080010103
Depth	100	100	100
Clay layer count	2	1	0
Clay total thickness	25	100	0
Clay thickest layer	15	100	0
Part-clay layer count	2	0	0
Part-clay total thickness	20	0	0
Part-clay thickest layer	15	0	0
Nonclay layer count	3	0	3
Nonclay total thickness	55	0	100
Nonclay thickest layer	30	0	40
Aquipro score	222	750	0
Clay in first 50 ft	15	50	0
Bedrock elevation	880	0	0
Drift thickness	70	0	0

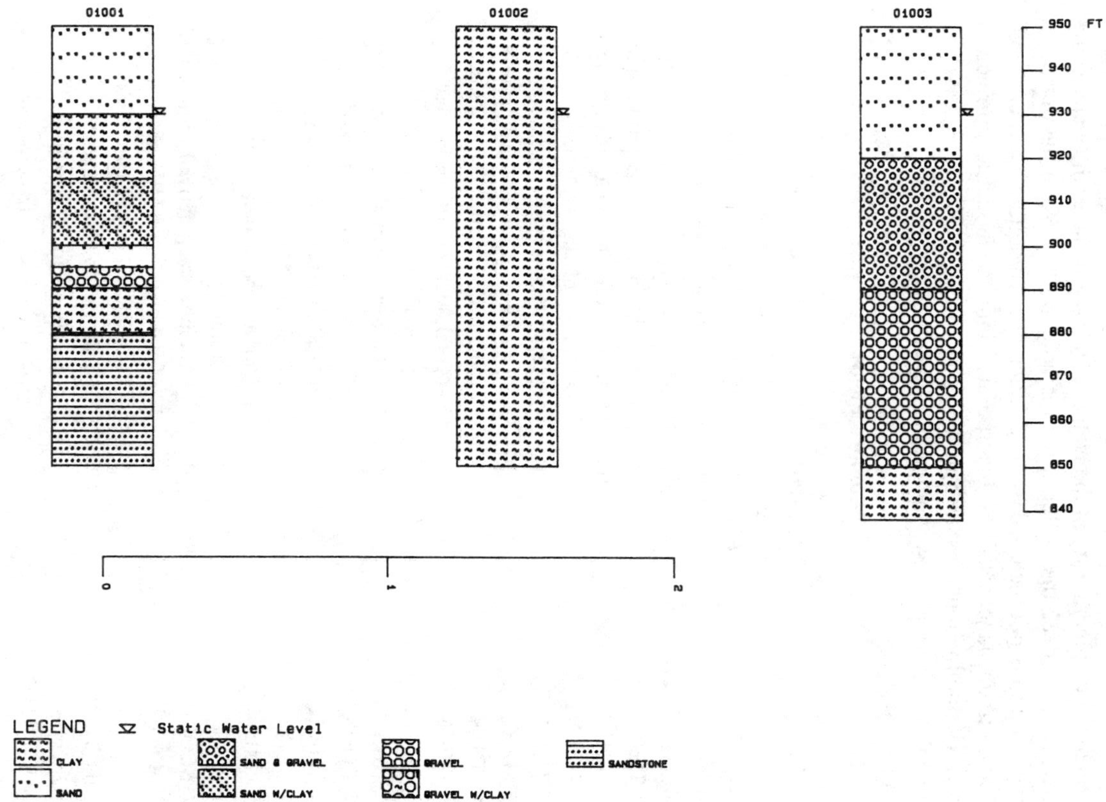

Figure 5. Hypothetical wells represent a range of Aquipro scores. All three wells are 100 ft deep. The first well is mixed lithology (score = 222), the middle well is all clay (score = 750), and the last well is all sand and gravel (score = 0).

The model takes into account specific circumstances. For example, bedrock is treated as a clay-type layer, unless it is the final layer, when it is treated as partial-clay on the assumption that it is permeable enough to serve as a source of groundwater. Also, if a screen is placed at some point above the well's depth, the bottom of the screen is used as well depth (as can be seen in the third well) on the assumption that the vulnerability of the well depends on the layer the water comes from more than how deep the driller bored the hole. In the case of the third well, the last layer (clay) is not included in the calculations. Undoubtedly, more modifications and circumstances will be incorporated into the model as correlations with known contamination sites are run.

The complete file created by Aquipro is shown below:

Aquipro Fields

WELLID	13030401001	(Well's ID number)
LOC_E	1893238.00	(MI coord. system, ft. east)
LOC_N	269433.00	(MI coord. system, ft. north)
ELEV	960.00	(Elevation above sea level)
DEPTH	80.0	(Well depth)
V_DEPTH	2.5	(Weight value for Aquipro formula)*
SWL	949.00	(Static water level elevation)*
C_COUNT	0	(Number of clay layers)*
C_THICK	0	(Total thickness of clay layers)*
C_MAX	0	(Thickness of thickest clay layer)*
P_COUNT	1	(Number of partial-clay layers)*
P_THICK	10	(Total thickness of partial-clay layers)*
P_MAX	10	(Thickest of partial-clay layers)*
N_COUNT	2	(Number of nonclay layers)*
N_THICK	70	(Total thickness of nonclay layers)*
N_MAX	45	(Thickest of nonclay layers)*
SCORE	25	(Aquipro score derived from formula)*
V_SCORE	1	(Aquipro group derived from score)*
CLAY_IN50	0	(Total clay in top 50 ft of well;* any interval can be specified)
BED_TOP	925	(Bedrock top elevation)*
AQUIFER	A1	(Aquifer, if found in well file)
DRIFT_THK	35	(Drift thickness, if bedrock was found)*
LOCAL_C		(Local code — an optional field)
LOCAL_N	41	(Local number — an optional field)
NOTE	0/0/ 80	(Local note — optional field)
CHARHEIGHT	200.0	(Character height default for mapping)*

CHARWIDTH	200.0	(Character width default for mapping)*
CHARANGLE	0.0	(Character angle default for mapping)*

Fields marked with * are calculated by Aquipro; the rest of the fields come directly from the well data file.

Aquipro scores are used to draw inferences about the vulnerability of the well based on lithology and well depth. Other factors that influence vulnerability are not included in Aquipro's analysis because the information is not available in the database. Other factors can be added as data become available; for example, aquifer vulnerability is also affected by well construction, nearby sources of pollution, land use, soil types, topography, and recharge. Drastic is an aquifer vulnerability model that uses:[3]

D	Depth to groundwater
R	Net recharge
A	Aquifer media
S	Soil media
T	Topography
I	Impact of the vadose zone
C	Hydraulic conductivity of the aquifer

The actual Aquipro formula takes these variables into account: well depth, amount of clay, and partial-clay material found in the lithology, as well as the thickness of the individual clay and partial-clay layers encountered (e.g., one clay layer 20 ft thick provides more protection than 20 clay layers 1 ft thick). The program does produce other useful information from the well data, including drift thickness, amount of clay in the top 50 feet of soil, and the bedrock elevation.

Aquipro's analysis can be useful in deciding which wells to sample for a chemistry database. For example, wells could be grouped as bedrock or non-bedrock (drift) based on values in the bedrock elevation field (Bed_top); wells with low Aquipro scores could be examined to evaluate their potentiality as aquifer recharge areas. The elevation, static water level, and bedrock could be contoured to create a map similar to Figure 6.

6. SUMMARY OF TOOLS FOR STUDYING GEOLOGY OF THE DATABASE

At this point in developing a groundwater database, the only data available on the computer comes from water well construction records. These data are adequate to help draw a preliminary picture of the hydrogeology of the region. A more comprehensive picture will be drawn when the next phase of the groundwater database is complete, the groundwater quality database. For now, data on water

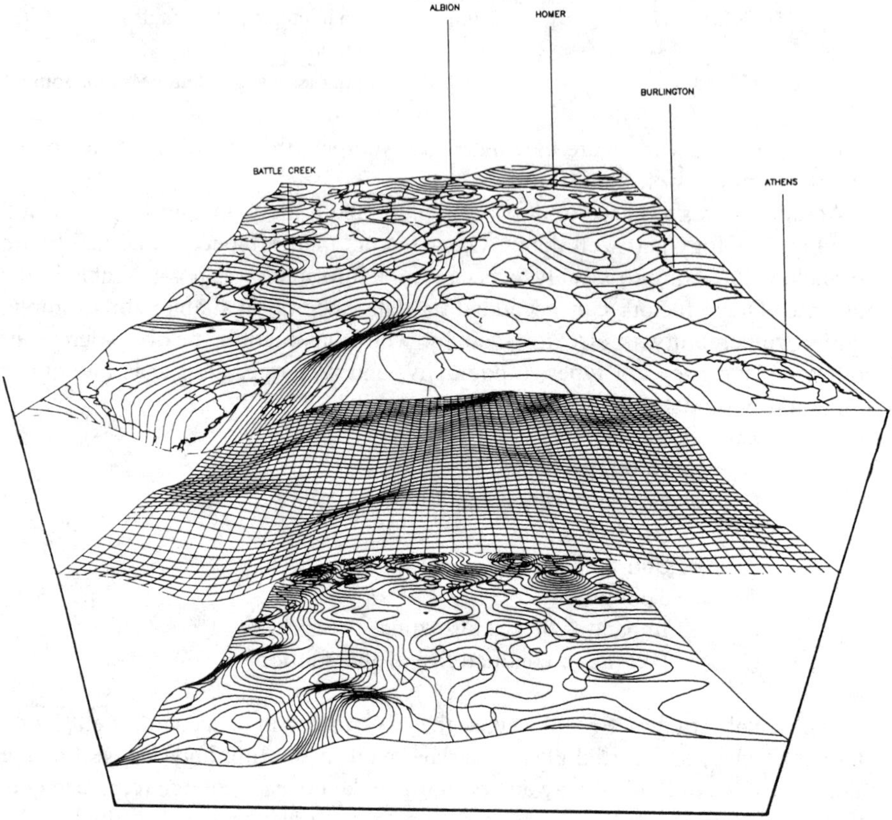

Figure 6. Contour map shows elevation, static water level and bedrock layers for Calhoun County, Michigan. (From the Institute of Water Science, Western Michigan University, Kalamazoo.)

quality, geology of the region, and surface features are limited to what is available from other sources such as the U.S. Geological Survey, state health department, or state natural resource agencies.

Included in the data mentioned above are the cross-sections describing the general lithology of an area, the static water contour map showing groundwater flow patterns, and a aquifer vulnerability map (Figure 7), indicating sites sensitive to groundwater pollution. The cross-sections show that a majority of the wells are using a common bedrock aquifer (Figure 3) which past reports[2,5-9] identified as the Saginaw Bedrock formation. Some well construction records indicate the use of some shallow glacial drift aquifers, but the primary water source is the bedrock aquifer. Lithology information from the cross-sections in conjunction with surface data on the topographic map also reveal possible areas of hydrogeologic connection between the ground surface and the bedrock. Some of these areas may be

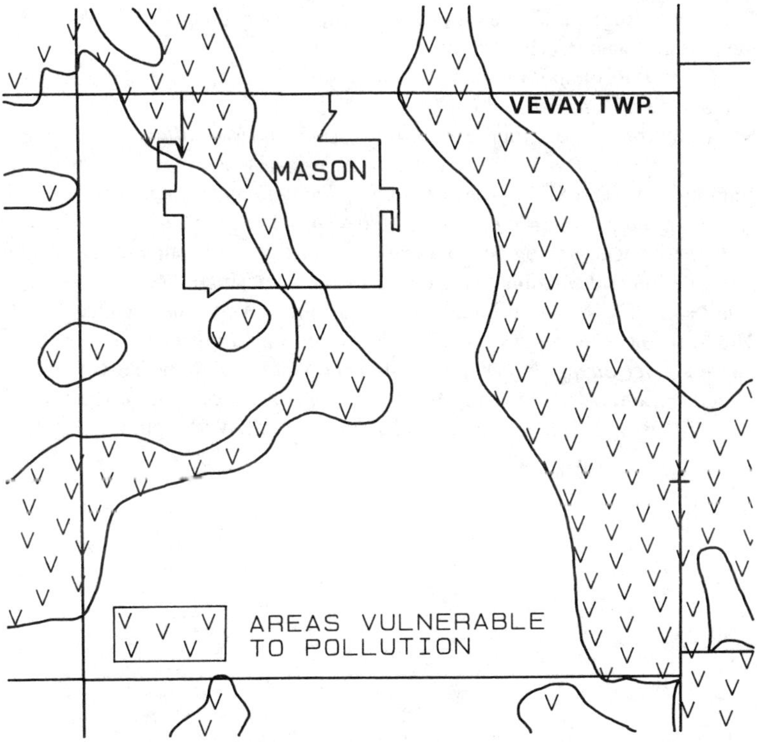

Figure 7. A groundwater vulnerability map indicating areas in Vevay Township where information from geologic cross-sections, static water level contours, and the Aquipro program suggest that aquifers in these areas are more vulnerable to pollution than others. (From the Ingham County Health Department, Michigan.)

groundwater recharge zones or locations where surface water is infiltrating downward to recharge the bedrock aquifer.

REFERENCES

1. Geobase4, Cross-section software by Hall Groundwater Consultants, Earthware of California, 30100 Town Center Drive, Suite 0196, Laguna Niguel, CA 92677 (714)495-5727.
2. Rowe, G. *Report on the Aquifers of Ingham County: A Survey of Groundwater Quality in the Saginaw Formation for Ingham County* (1989).
3. Surfer4 Software, Golden Software, Inc. P. O. Box 281, Golden, CO 80402.
4. Passero, R. M., F. J. Cohen, S. J. Dulaney, P. M. Haff, and H. Moaddel. "Aquifer Vulnerability Mapping in Southwest Michigan," *Policy, Planning and Resource Protection- A Groundwater Conference for the Midwest: Proceedings* (Kalamazoo: Western Michigan University, 1988), pp. 29-50.

5. Tri-County Regional Planning Commission, Ingham County Solid Waste Management Plan, Lansing (1983).
6. Tri-County Regional Planning Commission, Metropolitan Area Groundwater Management Plan, Lansing (1977).
7. Snell Environmental Group, Inc., *Groundwater Investigation: A report for the Tri-County Regional Planning Commission,* Lansing (1977).
8. Vanlier, K. E., and M. L. Wheeler. *Analog Simulation of Groundwater Development of the Saginaw Formation Lansing Metropolitan Area, Michigan,* prepared by the U.S. Geological Survey in cooperation with the Tri-County Regional Planning Commission and Michigan Geological Survey, Lansing (1968).
9. Vanlier, K. E., W. W. Wood, and J. D. Brunett. *Water Supply Development and Management Alternatives for Clinton, Eaton, and Ingham Counties. Geological Survey Water-Supply Paper,* prepared in cooperation with the Tri-County Regional Planning Commission and the Michigan Department of Natural Resources, Stock No. 2401-02442, U.S. Government Printing Office, Washington, D.C. (1969).

CHAPTER 7

The Water Chemistry Database – Measuring Water Quality

Complete groundwater studies have both a physical and a chemical component. As described earlier, the well construction records help describe the physical components that comprise the "world" in which groundwater resides and moves. A water sample and chemical analysis helps to describe the chemical component of the groundwater, which is also a reflection of the physical component.

As an integral part of the hydrologic cycle, groundwater is replenished from various sources of surface origin. Water from rain and snow moves into the ground and through a variety of geologic materials, picking up elements that will determine its chemical characteristics. Other components may also be taken up by water from the air before it moves into the soil. When water takes on a chemical characteristic, it may become unsuitable or suitable depending on the specific use. All groundwater becomes contaminated, meaning there is something in or with the water other than pure H_2O. This is different from the term "polluted," which for this discussion means that the water contains something that renders the water nonusable, depending on the specific need for the water. Pollution may come from both natural sources such as brine deposits, or from unnatural sources such as gasoline from a buried storage tank or salt from a road salt storage facility.

1. SELECTING THE SITES FOR SAMPLING

The work previously done out of the well construction database and other available research should generate some ideas about the general components of the geology for the study area. This includes the location of possible groundwater recharge and discharge, aquifer types, and locations.

At this point in groundwater database development and the creation of the water chemistry segment, one must decide whether the database will focus on those areas known to be polluted, naturally or unnaturally, or on areas which should reflect "background" water quality. The term "background" can be ambiguous when referring to water quality. Does it mean the quality of the water before being affected by any actions of humankind, or before the water is considered polluted, based on the specific needs of any one user? Most people would like to use it in terms of the effect humankind has had on the environment and on groundwater. Most of the wells sampled for the database would include the wells mapped for the well-log database. Because of this, it would seem difficult to sample for water which to some degree has not been influenced by any of the actions of humankind. Most wells are usually near sites that have been influenced by human actions.

For the purpose of this discussion it is assumed that the database user will want to focus on sampling "background" groundwater, defined more or less as water that has not been significantly affected by human activity. More specifically, this means water quality that has not been rendered nonusable for drinking water purposes. Using this as a standard to begin the selection process, the database user must avoid selecting any wells within a specified distance from a known or suspected source of groundwater pollution. For many sites this should be $^1/_2$ mi in any direction from the known or suspected pollution site. This $^1/_2$ mi distance comes from a review of several files at the Michigan Department of Natural Resources of known groundwater polluted sites and the distance pollution plumes show their influence.

Another factor in the water chemistry database development is the economics, or the amount of funds available for sampling. Both inorganic and organic analyses of the water can be expensive and money will be the most influential factor. Idealistically, one may want to sample the groundwater for everything possible, but this is usually not practical since many of the things that are possible to test for in water may not even apply in the particular study area.

Some of the groundwater study tools (discussed later) use some of the major cations and anions commonly found in most groundwater. These include calcium, magnesium, sodium, chloride, iron, potassium, sulfate, bicarbonate, and carbonate. Other parameters usually tested for include nitrates, hardness, conductivity, fluoride, pH, and total dissolved solids. While it might not be feasible to sample for more parameters than these, it would be recommended that these 15 mentioned be considered as a minimum to sample for in the chemical database. Again the database user must determine those parameters to be most important for sampling since the limiting factor is always the budget.

In connection with public health, bacteriological testing should also be considered to check for the sanitary conditions of both the well and distribution system. This usually involves sampling for the presence of coliform bacteria which originate in the intestinal tracts of humans and other warm-blooded animals. This test can help determine if the well is being impacted by some source of pollution, such as a septic tank or a sewage system. If such a well is found, steps must be taken to establish the source and correct the condition. If a well is so compromised, the well may not be suitable for the sampling survey, at least not for one focused on "background" water quality.

The database user must also do some laboratory service shopping to determine which lab will perform the work based on the certification, reputation, and prices of each laboratory. The location of the lab is also a factor because of the time constraints involved with simple delivery.

For any given study area the goal should be to characterize the water chemistry in an identified aquifer as the groundwater moves from areas of recharge, through transition zones, and out through areas of groundwater discharge. The idea is to observe the changes in water chemistry throughout the entire aquifer system. For a given area there may be more than one aquifer system involved, therefore, the researcher must decide if all the aquifer systems or perhaps the one system most often used will be studied . Again, economics usually play the most important role in these kinds of decisions.

A simple illustration is shown in Figure 1. In this scenario for Delhi Township in Ingham County, Michigan, areas of possible groundwater recharge were identified with discharge possibly occurring in locations where streams or ponds are located. Unmarked areas in between may be considered transition zones. It should be reemphasized that the determination of groundwater recharge or discharge is difficult, and usually requires detailed on-site work; however, some general ideas of the location of these areas can initially be posited through the use of both the well construction database and any previous research. A majority of the well construction records indicate that the primary aquifer used is a bedrock aquifer, which in this case has been called the Saginaw Formation by geologists doing work in this area. A static water contour map (Figure 2) gives a general picture of groundwater flow for the study area.

Assuming a limited budget, only 20 wells can be sampled for an area the size of a township, so in the Delhi example the 20 wells must be spread between the recharge, discharge, and transition areas. These wells all represent the one bedrock aquifer. In practice, not all the wells chosen are picked for best representation of an area, but rather where well owners allow sampling. Also, no wells may exist in a particular area a company or individual wishes to represent. Having these limitations the wells picked for sampling are shown in Figure 1. It is hoped that these wells will represent water chemistry in the three geologic zones, taking into account groundwater flow direction.

Along with selecting the wells locationally to get the desired samples, other practical matters must be considered. These considerations should include:

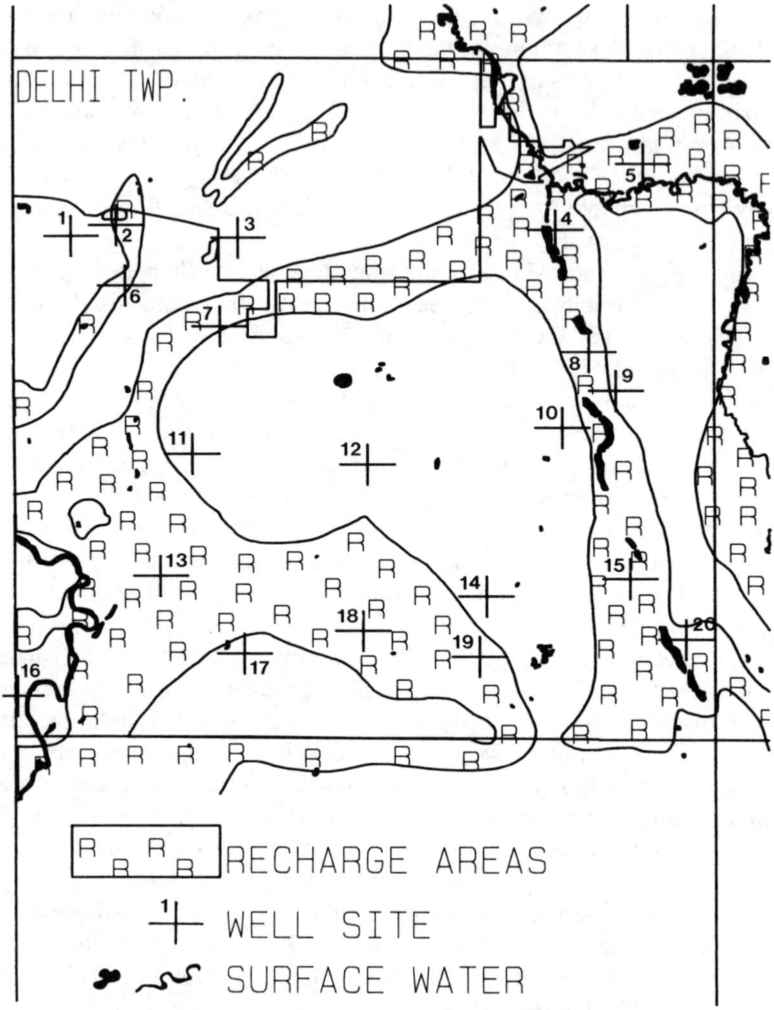

Figure 1. Map of Delhi Township, Ingham County, Michigan. Possible groundwater recharge zones are shown as labeled. Possible discharge areas are assumed to be in locations in which surface water such as ponds, streams, etc., are found. Unmarked areas are considered the transition zones. Wells 1 through 20 are the chosen wells for taking samples. Often the best locations for well sampling are not available because well owners will not allow their well to be sampled or the wells are not located in the best areas.

1. Cooperation of a homeowner, facility owner, etc. to participate in the sampling project, especially if the project involves sampling over several years.
2. Documentation for the well should be complete, especially for the well construction record and/or a permit for the well.

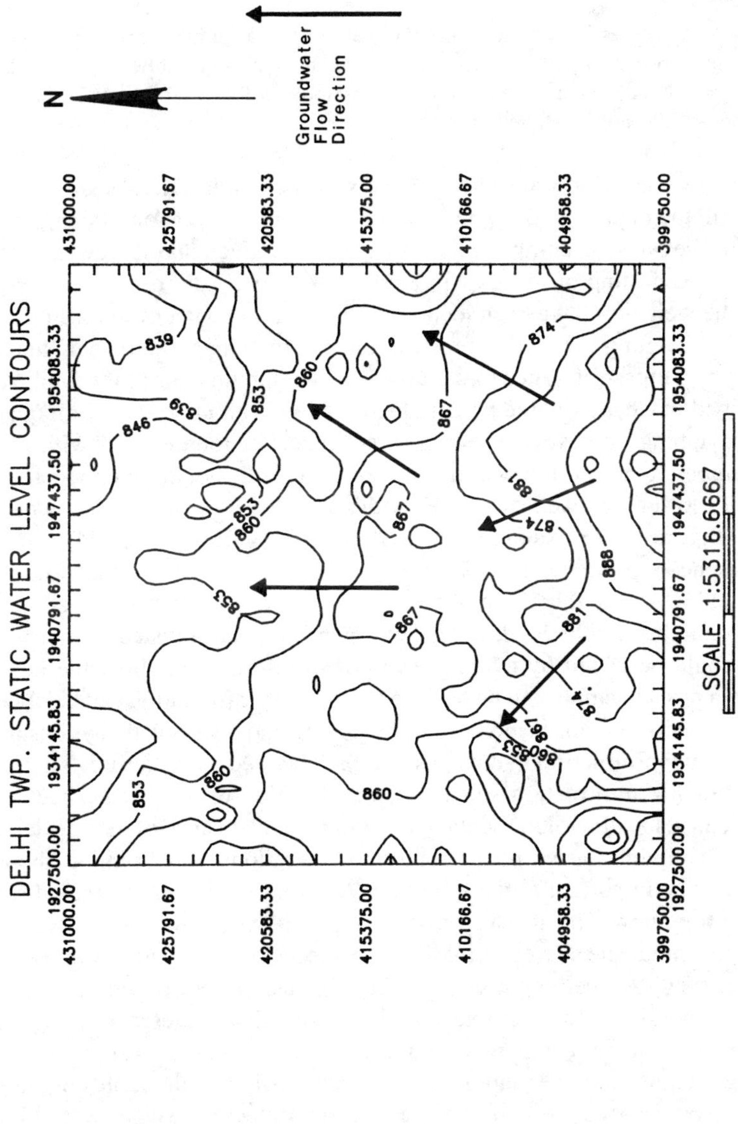

Figure 2. Static water level contour map of Delhi Township showing general groundwater flow in the bedrock aquifer.

3. Can "raw" water be sampled from the well, or can a sample be obtained that does not go through water treatment devices such as a water softener, iron filter or distillation unit? Because a representative sample of the groundwater is needed, the water should not be altered in any way.

With a few exceptions, most homeowners, owners of businesses, etc., are very interested in the quality of the water they use. It is proper that once sampling and analysis of the water is complete that the results be given and reviewed with the well owner. Reporting of results will be discussed in greater detail in Chapter 8.

Once the well sites have been identified, the current owner of the well must be identified. Assuming the well is properly located on a map, the owner should be identified by a cross-reference to the tax identification number of the land parcel. As described in Chapter 6, the parcel identification is an important component of the well-mapping process. Once the correct parcel is found along with the tax parcel number, a cross reference to an updated tax roll should give the name and address of the site and well owner. How this tax parcel system works is demonstrated in Figure 3. For example, well 1 shown in Figure 1 (upper left corner) is cross-referenced to the tax map (Figure 4). Once the site parcel is identified, the current landowner is found on the tax roll. Using the name and address, a telephone number is obtained, unless the owner has an unlisted number. Well owners should be called by telephone to obtain information about the home or facility so a proper sample can be obtained. The same information can be obtained through the mails, but this is more time consuming and does not always ensure the more complete information that can usually be obtained by directly talking to the well owner. Site information should include location of any buried fuel oil or chemical tanks, location of the septic tank or drainfield, history of the site, location of the best sampling faucet, possible farming operations, and anything about the site that may be important to the sampling. A survey form can be created for the researcher to review with the well owner.

After this initial site survey, it may become necessary to delete a proposed well from the survey if something about the site suggests that a representative sample may not be possible. An example may be a well that serves a Department of Transportation salt storage facility, with the well located only 25 ft from the bulk salt storage. The well owner should be made aware of the vulnerable condition of the well so some corrective steps may be taken to protect the well or water supply; however, such a site would not be ideal if the goal is to look at background water quality.

As the well owners are surveyed by phone, appointment times and arrangements for well sampling can be made. Within an 8-hr workday, two people working in tandem can sample ten wells or well sites. This includes completing a site survey, checking the plumbing to confirm an appropriate sampling faucet, running the water to get a fresh sample from the well, and collecting the samples. If conductivity, pH, and water temperature are collected and if a budget is allocated, these parameters should be collected with an instrument on-site. There

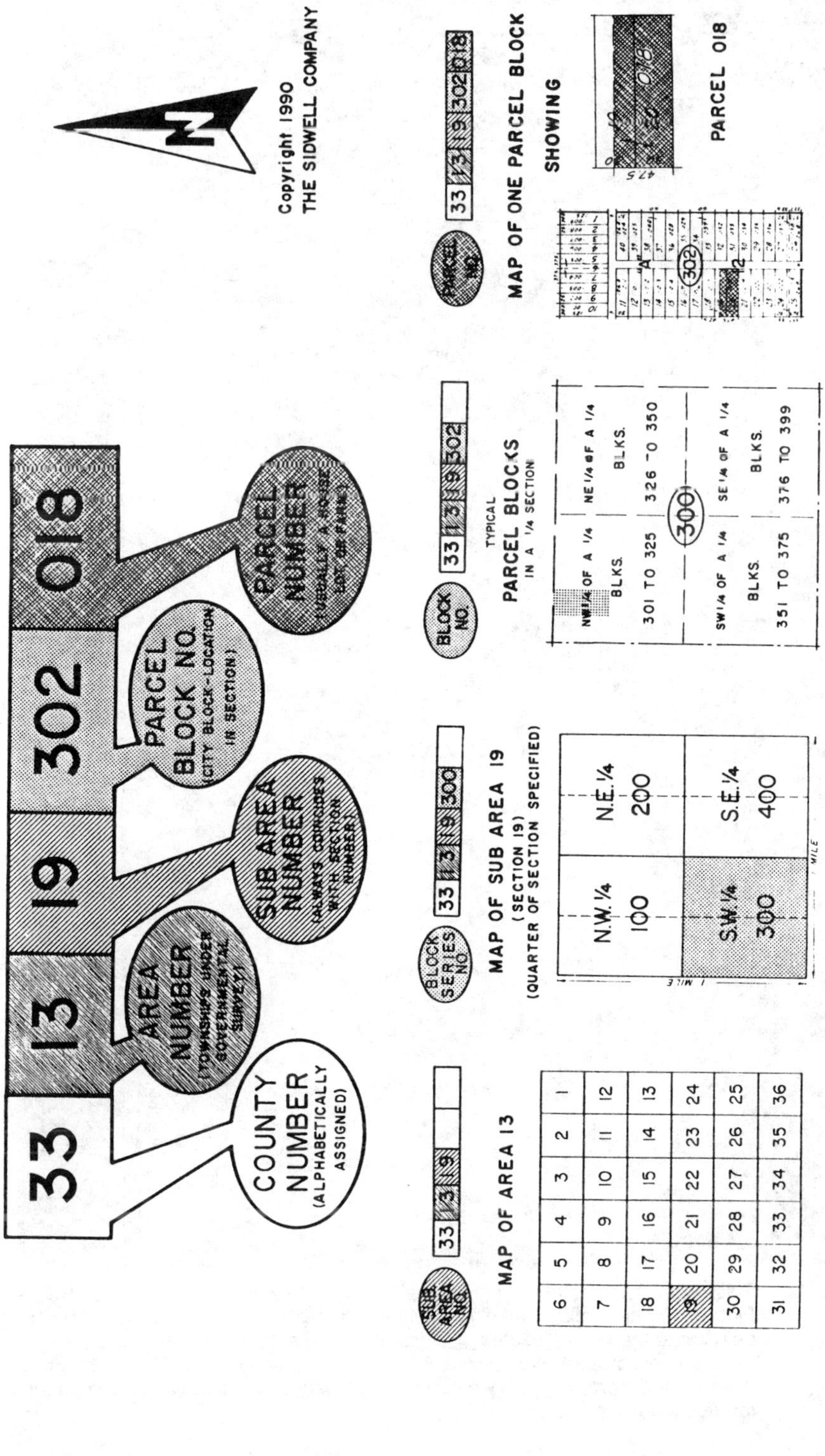

Figure 3. A graphic illustration of how the tax parcel number system works. Ingham County has a County number of 33. (From The Sidwell Company, Chicago.)

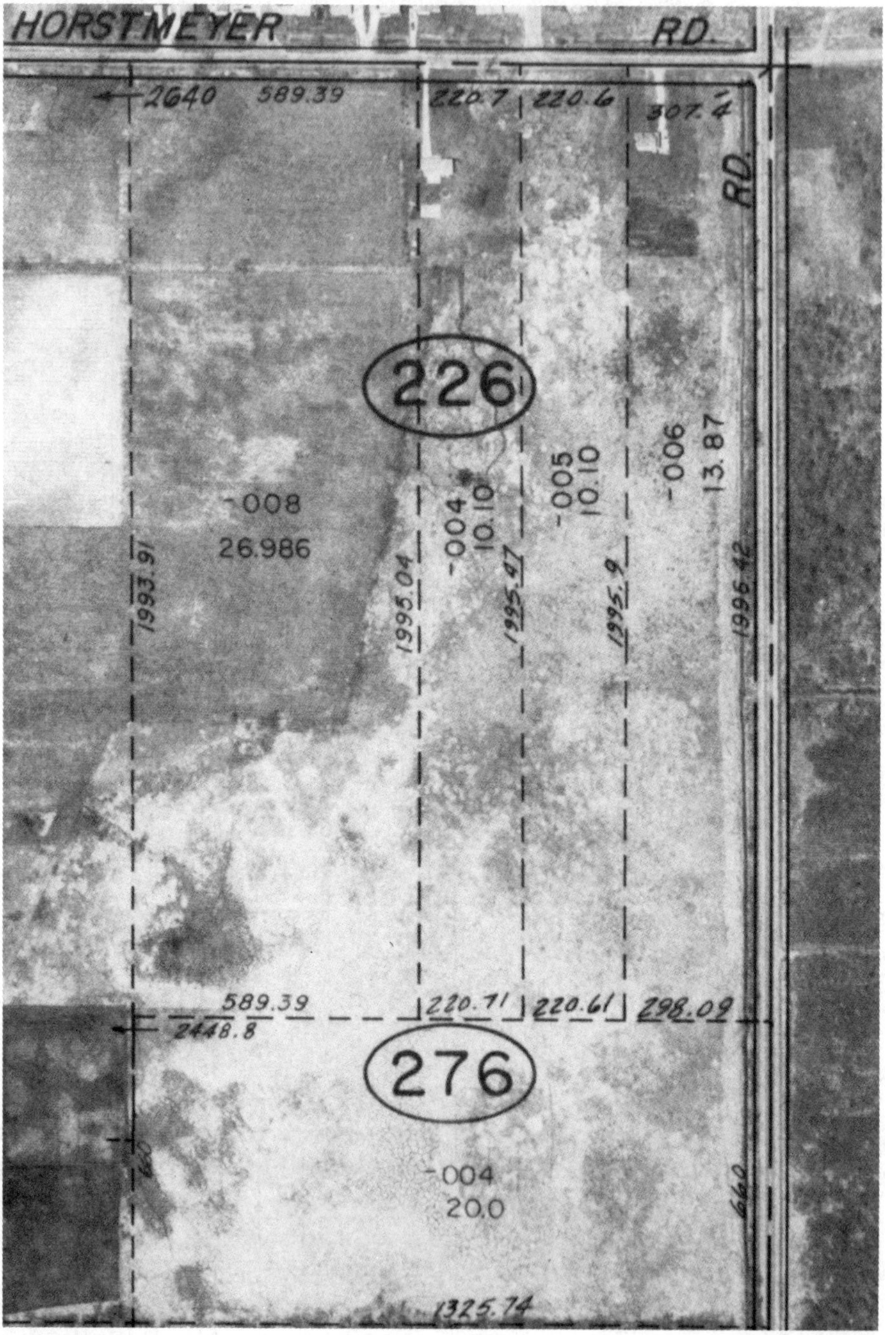

Figure 4. This Aerial-Tax parcel map gives the location of sampling well number 1 from Figure 1. The well, located in Delhi Township (Township 05), Section 18, belongs to the house located in Block 226, Parcel 006. From the illustration in Figure 3, this property would have a tax parcel number of 330518226006. From this number, a property and well owner can be contacted to make arrangements for well sampling.

are several companies that supply on-site conductivity, pH, and temperature instruments (see Appendix D). Once at the site, the water should be run long enough to flush the water system of any "standing" water so a fresh or representative water sample is collected from the study aquifer.

The standard method for sampling monitoring wells established by the U.S. Environmental Protection Agency (EPA) calls for the total volume of water in the well to be flushed three times. This is reasonable for monitoring wells since these wells are infrequently used, sampled perhaps only once a year. Standing water in such a well would stagnate in the well casing which may change the water chemistry and would not provide a representative sample of the groundwater.

If a water supply well for a home or business is used in the survey, a different condition exists because the well is usually in operation every day. In this situation the water should not stagnate in the well or building plumbing very long, provided the home or facility is not vacant. Under normal conditions in which a home or building is occupied, and the time of sampling follows a high level of water use for that particular day, a representative sample may be obtained if the well and plumbing is flushed by turning on two or three faucets for approximately 20 min. This is after the plumbing system has also been checked for filters, treatment devices, and faucets that are nearest the well in pipe distance. After running the water, another check run by the sampler measures the water temperature and evaluates whether it reflects what may be expected for groundwater. For Ingham County, Michigan, this may be between 9.5° and 14.5°C.

For a sample less open to variables, the procedure for sampling a "monitoring" well may be followed.[1] Standard practice is to remove (purge) three times the volume of water held in all the well and plumbing systems involved. Some standard pipe sizes have these gallons per foot:

$^1/_2$-in. pipe	=	0.010 gal/ft
$^3/_4$-in. pipe	=	0.023 gal/ft
1-in. pipe	=	0.041 gal/ft
$1^1/_4$-in. pipe	=	0.064 gal/ft
$1^1/_2$-in. pipe	=	0.918 gal/ft
2-in. pipe	=	0.163 gal/ft
4-in. pipe	=	0.653 gal/ft
6-in. pipe	=	1.469 gal/ft
8-in. pipe	=	2.611 gal/ft
10-in. pipe	=	4.080 gal/ft

For example, with a 60-ft well and all pipes:

50 ft of 4-in. casing	50 ft × 0.653 gal/ft	= 32.65 gal
10 ft static water level		
Riser pipe of 1-in. diameter	10 ft × 0.041 gal/ft	= 0.41 gal
25 ft, 1 in. to residence	25 ft × 0.041 gal/ft	= 1.03 gal
25 ft, $^3/_4$ in. inside plumbing	25 ft × 0.024gal/ft	= 0.58 gal

40 gal pressure tank	40 gal/2	= 20.00 gal
Total system volume		= 54.67 gal

If the faucet used has a flow rate of 3 gal/min, (3 GPM), 55 min of pumping will purge the system of three times the volume in it. To minimize this time, all cold water taps can be run concurrently. An outdoor hose tap typically flows about 5 to 7 GPM, as often does a bathtub or shower tap. By also opening the bathtub tap and one outside faucet, the system could be purged in 9 min, assuming the pump can maintain an 18-GPM flow rate. The sampling tap is allowed to run for a few minutes longer than the others before a sample is taken.

The aeration filter should be removed from the faucet. When filling the bottles, the sampler should decrease the flow from the faucet so a small but steady stream of water is discharged, since some bottles may contain an acid preservative such as nitric acid for heavy metals. These steps will also prevent air from getting into the sample, which is especially important if volatile organic sampling is done using a special bottle and cap. The samples should also be kept cool, preferable in a cooler with ice, and delivered to the laboratory as soon as possible. Before leaving the home, the well owner should be informed as to the date he/she may expect the results and be given the sampler's telephone number should the well owner have questions.

2. REVIEW OF THE DATA

After sample results have been received from the laboratory, they should be screened for any potential public health problem, including the presence of coliform bacteria or some parameter exceeding a drinking water standard or MCL (maximum contaminant level, established by the U.S. EPA). The first priority is the protection of the people using the water and the aquifer system.

If a parameter does indicate some public health problem, the well owner should be notified immediately and advised how to get at least a temporary safe water supply, such as bottled water from an approved source. Other regulatory agencies may be notified, especially for a severe problem of groundwater pollution. During this time, arrangements must be made to resample the well to confirm the problem. This may also include a call to the laboratory to review the results. Most laboratories will contact the project leaders with a suspicious result, but it is prudent to make arrangements for such contact with the selected laboratories. After the data have been screened for any parameters of public health concern, other general patterns in the data should be reviewed. Much depends on the current information available for the area sampled and expectations. Does the area normally have a high salt content? Is the water normally hard? Are there typically odor and taste problems associated with iron or manganese? If there is no information, or "no one person's" experience available for review, the samples taken for the survey may be innovative.

A quick, preliminary review of the data will (hopefully) show any discrepancies that may be attributed to a laboratory error. It is much easier to uncover problem data at this point rather than after the data are entered into the database where it could be forgotten. Also, it is easier to correct a problem if some samples are still held by the laboratory.

When this review of the results is complete, the data can be arranged for entry into the computer's water chemistry database. One of the first steps may involve sending a letter and summary of the results to the homeowner who allowed the sample to be taken. When the data are in a database, a simple database report can be used or the data can be merged into a word processing program. The homeowner is interested in the test results, and will appreciate any effort to communicate with them. Having a database with addresses and chemistry results facilitates this communication process.

Once in the computer, the data can be further analyzed and manipulated with various software systems. In Chapter 11, a number of practical examples are described using the chemistry database in conjunction with various software tools.

REFERENCES

1. *Michigan Groundwater Survey Users' Guide* (Kalamazoo: Western Michigan University, 1989), p. 11-1. Science for Citizens Center, WMU, Kalamazoo, MI 49001.

CHAPTER 8

Entry of Chemistry Results – Electronic Files

The very water chemistry results that are keyed into a local database have probably been keyed into other computers at the laboratory. In the future, the EPA STORET[1] system may provide a common storage bank for surface and ground-water quality data, and test results are distributed electronically. It is also likely that companies will continue to maintain their own versions of databases on microcomputers, by downloading or typing test results into water chemistry files.

1. COMMON ISSUES IN GROUNDWATER CHEMISTRY DATABASES

Repeated Samples

Groundwater chemistry databases must take into account repeated sampling of the same site. Because of this, a simple well identification number is not enough to distinguish one sample from another. A combination of the well ID number and the sample episode must be devised. Further, samples may be sent to more than one laboratory, and in such a case, a three-field key consisting of ID, sample episode, and laboratory may be necessary. Database design and programming are more complicated because of the need for one-to-many file relationships (one well, many samples).

Detection Limits

Another issue is how to store information on the detection limit involved. Two common methods are

1. Simply use a minus sign to indicate "not detected at" the value shown. For example, if the detection limit for arsenic was 0.05 m/l, and no arsenic was detected, then the value entered for arsenic is -0.05. The table below is an example of a small data listing:

Well No.	Compound	Value
08072901	Arsenic	0.10
08072902	Arsenic	-0.05
08072903	Arsenic	0.05

2. Use two fields for each chemical result, one to report the value or ND (not detected), and the other field to report the applicable detection limit.

Well No.	Compound	Value	Detection Limit
08072901	Arsenic	0.10	0.05
08072902	Arsenic	ND	0.05
08072903	Arsenic	0.05	0.01

The advantage of the second method is that detection limits can be accurately tracked, since they will vary from laboratory to laboratory and project to project. Emphasis should be placed on the detection limits, particularly for statistical treatment of data. For example, consider this exaggeration: To compute the average arsenic for the samples above, one could arrive at four different values:

1. Throw the nondetect sample out (0.10 + 0.05)/2 = 0.0750
2. Use the detection limit value (0.10 + 0.05 + 0.05)/3 = 0.0667
3. Use half the detection limit value (0.10 + 0.025 + 0.05)/3 = 0.0583
4. Consider the nondetect as zero (0.10 + 0.00 + 0.05)/3 = 0.0500

2. GROUNDWATER CHEMISTRY MINIMUM DATA ELEMENTS

In the short run, a database can be limited to the exact project parameters desired by the organization. For example, the Michigan Groundwater Survey decided on a specific assortment of chemicals they wanted to test to establish background water quality.[2] These specific chemicals and a wellid to key the chemistry results to the well construction log constituted the database. When

```
===============================================================

        Michigan Groundwater Survey Partial CHEM File        v3.1

*******************************************************************************

LOCATION:  W.NO    330053603   (cctttssxx)    LAB1 999    LAB2        LAB3

        L.NAME  GARLOW                                  COL.DT  08/31/82

 *Values ppm except Con.(MMHOS) *Not Detected as -det.level *Not Tested as 0.00

    Alkalinity    337.000  <  Aluminum       .030      Ammonia        .000
  < Antimony         .000  <  Arsenic       -.050  <  Barium         .000
  < Boron            .000  <  Cadmium       -.010  <  Calcium      96.100
    Chloride        3.000  <  Cobalt         .010  <  Chromium      -.010
    Conductivity  640.000  <  Copper        -.010     Fluoride       .100
    Hardness      332.000  <  Iron          4.700     Iron.2        2.222
  < Lead            -.050  <  Magnesium    24.700  <  Manganese      .140
  < Mercury         -.050  <  Molybdenum    -.010  <  Nickel        -.010
    Nitrate         -.300     pH            6.500  <  Phosphorus    -.100
    Potassium        .900  <  Selenium      -.050     Silica         .000
    Sodium          4.400     Sodium.2      4.440     Sulfate      62.000
    Detergents       .000  <  Thallium      -.050  <  Zinc          -.010
    TDS           384.000     C        V     .000     C1     V1      .000

            Note1 SITE 6C AL              Note2 MDPH auto

===============================================================
```

Figure 1. Computer entry screen for Michigan Groundwater Survey of background water chemistry in southwest Michigan. (From Science for Citizens' Center, Western Michigan University, Kalamazoo, 1989.)

chemicals were not detected, a minus value was used. An example of such a database is shown in Figure 1.

In the long run, groundwater chemistry data are more easily aggregated when a common file format is adhered to. The EPA STORET system is likely to become

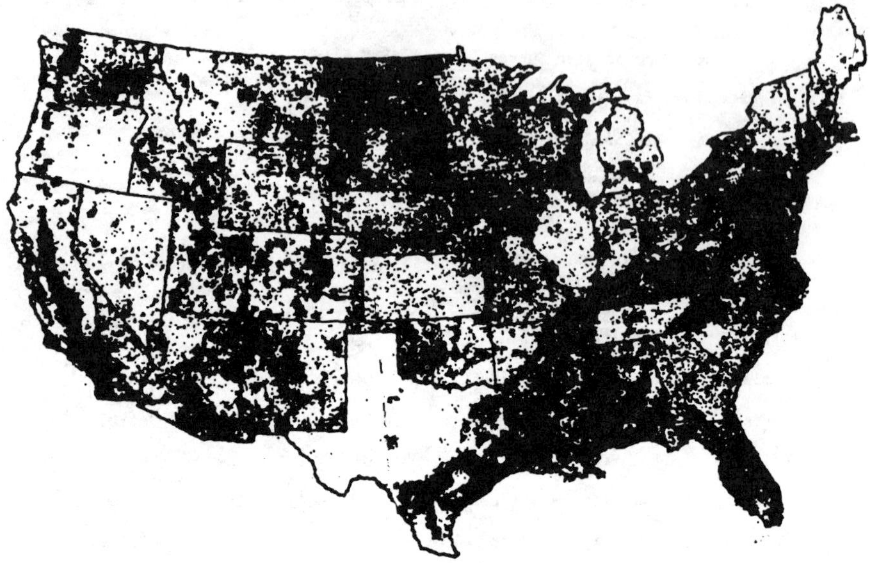

Figure 2. Extent of STORET coverage including WATSTORE sites. (From EPA 440-6-87-005, Office of Groundwater Protection, May 1987.)

the standard for shared groundwater chemistry data. The STORET system has many fields and codes. Its dimensions in 1987 were "over 9500 unique water quality parameters in STORET Water Quality File, and it contains over 90,000,000 observations on samples taken from more than 600,000 sampling stations. Of these Water Quality Monitoring stations in the Water Quality File, 270,000 are Ground-Water Monitoring stations located throughout much of the United States. . . ." The 270,000 stations also include the U.S. Geological Survey's 160,000 Ground-Water Quality Monitoring Stations (WATSTORE). Most of the STORET data was collected from 1975 to the present. Figure 2 shows the extent of STORET coverage with WATSTORE.

The STORET database accommodates most information. An example of its thoroughness is the following listing of the station descriptors that pertain to groundwater users:[1]

Facility name
Type of waste management area (landfill, etc)
RCRA regulating agency
Facility Dun & Bradstreet number
State/county codes and names
Major/minor/subbasin codes and names
Hydrologic unit (USGS)
Facility type (treator, storer, etc.)

Other permits held by facility
Permit status
Ownership of facility
Legislation of concern
Facility history
Land use
Potential targets, nearby sensitive ecosystems
Contacts for further information
Total well field description
Other monitoring activities
Latitude/longitude
Well "position" relative to waste management area
Purpose of well
Type of monitoring
Drilling method
Screened interval
Depth of hole
Depth of well
Well intake finish
Well casing material
Type of material from which opening is made
Proximity and position relative to drinking water sources
Lithologic modifier
Source of geohydrologic data
Topography
Geologic age or formation name
Aquifer name
Lithology
Aquifer thickness
Depth to top of lithologic unit
Depth to bottom of lithologic unit
Drilling fluid
Available logs
Type of surface seal
Bottom of seal
Method of development
Top of casing below LSD
Bottom of casing segment below LSD
Casing thickness
Diameter of open section
Width of opening
Length of well opening
Source of evacuation data
Casing inner diameter

Casing outer diameter
Bore hole diameter
Packing material
Date of analysis
Date of extraction
Type of water use

Each of these parameters has detailed codes. For example, the codes available for topography are

Value	Value Description
VALLY	Valley
SLOPE	Slope
HLLTP	Hilltop
DRAW	Draw
FLAT	Flat
ALVFN	Alluvial fan
PLAYA	Playa
STRCH	Stream channel
LCDEP	Local depression
DUNES	Dunes
FLPLN	Flood plain
SNKHL	Sinkhole
LSM	Lake, swamp, or marsh
MNGRV	Mangrove swamp
ESTRY	Offshore (estuary)
PEDMT	Pediment
TERRC	Alluvial or marine terrace
UNDLG	Undulating
OTHER	Other

The Sample Process

Similarly, each sample activity can be entered in detail. The following is a listing of the sample descriptors:

Volume of water evacuated from well prior to sample collection
Method of evacuation
Method of water level measurement
Monitoring well sampling method
Lab identification number
Sample number and type

Sampling and analytical technique
Pumping period
Sampling point description
Data quality assessment
Source of depth data
Method of depth measurement
Performer of sampling
Sampling type and frequency
Inspection type
Regulating agency
Sampling purpose
Production level
Sampling date
Elevation of land surface datum
Elevation of ground water
Depth to water level
Water level relative to mean sea level
Measuring point elevation
Depth from level to measuring point
Source of evacuation data
Source of water level data

They also are quite thorough in the codes for each descriptor. For example, the sampling purpose descriptor has the following codes:

BKGRD	Sampled to determine background levels
GWQAL	Sampled to determine ground-water quality
DWQAK	Sampled to determine ground-water suitability as a drinking water source
CNTMN	Sampled to determine groundwater contamination
ASMNT	Sampled as part of facility's assessment plan
PRMIT	Sampled as part of facility's permit requirements

Other Methods of Data Entry

If faced with data entry from massive amounts of paper reports, the purchase of scanners should be considered. Scanners and optical character recognition (OCR) software make it possible to electronically read groundwater chemistry results from laboratory reports into a data file. A scanner reduces the paper image to a pattern of dots, and OCR software converts the patterns of dots into letters and numbers. The speed and accuracy achievable with this technology is related to the quality of the original documents. Handwritten documents are usually not worth scanning, and computer or typewritten documents convert easily.

REFERENCES

1. Blake-Coleman, W., and N. Dee. *Ground-Water Data Management with STORET,* EPA 440-6-87-005, Office of Groundwater Protection, USEPA (May 1987).
2. *Michigan Groundwater Survey Users' Guide* (Kalamazoo: Western Michigan University, 1989),p. 11-1.

CHAPTER 9

Analysis of Chemistry Data – How to Interpret the Results

1. PROFILING THE WATER CHEMISTRY

An area of hydrogeologic study can sometimes take on a personality of its own to the observer. If enough information is available for an area the size of a township (for some townships approximately 576 mi^2), the database user may be able to locate a well by looking at the water sample results. With a groundwater database containing lithologic data, groundwater flow information, chemistry information, etc., a profile of the hydrogeology of an area can be made and studied.

In studying the human body, medicine offers several tools. X-rays can be taken to observe the inner structure and blood samples can be taken to understand the body chemistry. With the groundwater database, geologic cross-sections are made to observe the inner structure of the aquifer. Water samples can also be taken to study its chemistry. In order to include the chemistry profile, the database user will need the help of some tools, just as a doctor may need blood tests and X-rays. Chapter 6 discussed the use of one tool, the software from Hall Groundwater Consultants Inc.,[1] for creating geologic cross-sections. Other tools are available for researching the water chemistry. This chapter examines three general components of the water chemistry profile: statistics, graphics of values, and distribution geographically.

2. STATISTICS

Statistics can provide some very useful information as to the general trend of the data and what may be expected from future sampling of wells, such as:

1. The average concentration of any one parameter, such as calcium, for the study area
2. The variability of the water chemistry or its consistency
3. What kind of concentration ranges are expected
4. What correlations between certain parameters may be found, such as the correlation between water hardness and magnesium levels
5. Differences found in water quality both geographically or geologically
6. Differences found in water quality from suspected groundwater pollution problems and "normal" unaffected water

To proceed with many of the types of statistical analysis studies, a software package is used in conjunction with the water chemistry database. Some of the following examples of raw statistics were run by the Statistical Services Department at Western Michigan University. For some of the following examples, a software program called Statgraphics[2] was used.

Once the data are prepared for statistical analysis the original goals for creating the database should be kept in mind. In Chapter 7, some questions were asked by the database user before proceeding with the project:

1. Are we trying to determine "background" water quality only?
2. Are we creating a database of well water at sites of known, suspected, or potential groundwater pollution?
3. Are we only concerned about certain parameters of water quality such as organic pollutants?
4. Do we want the database to include anything and everything?

Based on the original intent of the groundwater survey and database, the way data are managed and categorized before use in a statistical program may be critical. For example, if the original goal of the database user was to establish "background" water quality, then some review of the data are needed. If organics were included in the well testing, then it may be necessary to exclude any wells that appear to be "polluted" from the statistical analysis. If the goal is to examine only background water quality, a "polluted" well may not reflect background levels for several parameters. Inorganic parameters may also reflect a polluted well such as elevated chlorides, sodium, sulfates, or iron. There are some techniques such as the use of STIFF diagrams which may point out a well that does not seem to conform to the norm. STIFF diagrams are discussed in the following Section 3, "Graphic Profiles."

For the purpose of this discussion, assume that the database user sampled wells in a survey with the goal of looking at "background" groundwater quality for an area the size of the county. The budget allowed for a total of 312 wells to be sampled for 33 inorganic parameters and a limited number of wells to be sampled

Table 1. Some Basic Statistics Run on the Survey of 312 Survey Wells: Groundwater Chemistry Analysis — Means and Standard Deviations for 312 Project Wells

County: INGHAM
Township(s): ALL
Lithology or Aquifer: ALL

	All Data			Detectable Concentrations Only				Nondetectable Concentrations	
Chemical	N	Mean	SD	N	%	Mean	SD	N	%
Alkalinity	312	286.7	56.4	312	100.0	286.7	56.4	0	0.0
Aluminum	•	•	•	14	4.5	0.094	0.130	298	95.5
Ammonia	311	0.172	0.151	298	95.8	0.179	0.151	13	4.2
Antimony	•	•	•	0	0.0	•	•	312	100.0
Arsenic	293	0.005	0.006	129	44.0	0.009	0.006	164	56.0
Barium	312	0.170	0.159	304	97.4	0.174	0.158	8	2.6
Boron	312	0.514	0.863	208	66.7	0.768	0.961	104	33.3
Cadmium	•	•	•	0	0.0	•	•	312	100.0
Calcium	312	79.9	44.7	310	99.4	80.4	44.4	2	0.6
Chloride	311	11.8	20.9	81	26.0	31.2	34.3	230	74.0
Cobalt	•	•	•	1	0.3	0.020	•	311	99.7
Chromium	•	•	•	0	0.0	•	•	312	100.0
Conductivity	312	647.5	211.0	312	100.0	647.5	211.0	0	0.0
Copper	•	•	•	37	11.9	0.037	0.055	275	88.1
Fluoride	303	0.5	0.4	301	99.3	0.5	0.4	2	0.7
Hardness	305	295.8	116.6	303	99.3	297.7	114.5	2	0.7
Iron	312	1.089	1.251	305	97.8	1.113	1.254	7	2.2
Lead	•	•	•	95	30.6	0.004	0.002	215	69.4
Magnesium	312	23.242	10.107	310	99.4	23.391	9.965	2	0.6
Manganese	312	0.042	0.084	287	92.0	0.045	0.086	25	8.0
Mercury	•	•	•	0	0.0	•	•	312	100.0
Molybdenum	•	•	•	1	0.3	0.010	•	311	99.7
Nickel	•	•	•	0	0.0	•	•	141	100.0
Nitrate	•	•	•	25	8.0	0.4	0.3	287	92.0
pH	•	•	•	300	100.0	7.148	0.539	0	0.0
Phosphorus	•	•	•	16	5.1	0.2	0.2	296	94.9
Potassium	312	3.803	2.709	312	100.0	3.803	2.709	0	0.0
Selenium	•	•	•	0	0.0	•	•	312	100.0
Silica	312	8.466	6.531	308	98.7	8.571	6.508	4	1.3
Sodium	312	26.837	46.312	312	100.0	26.837	46.312	0	0.0
Sulfate	312	46.7	81.5	301	96.5	48.3	82.5	11	3.5

Table 1 (Continued). Some Basic Statistics Run on the Survey of 312 Survey Wells: Groundwater Chemistry Analysis — Means and Standard Deviations for 312 Project Wells

	All Data			Detectable Concentrations Only				Nondetectable Concentrations	
Chemical	N	Mean	SD	N	%	Mean	SD	N	%
Thallium	•	•	•	0	0.0	•	•	312	100.0
Zinc	312	0.099	0.218	267	85.6	0.115	0.232	45	14.4

Note: For some parameters such as cadmium, no statistics were run because all the results were reported as "not detected."

[a] Nondetectable data are quantified at the midpoint between zero and the detection limit only for certain chemicals. The means for the other chemicals were not calculated and appear as a dot (•).

From the Ingham County Health Department, Michigan, analysis performed, Western Michigan University by Stoline and Miller.

for some selected organic parameters. Table 1 shows some preliminary statistics for inorganic parameters tested. Examination of the results gives an idea as to the nature of the water chemistry using simple statistics such as the mean and standard deviation. Another statistical tool is the percentile tables which describe the actual values of the chemistry results. Table 2 shows a percentile chart of 36 parameters created for this example. The 50% values should reflect the average values for each parameter. When sampling other wells for other studies or investigations, the results of these other tests can be compared to the percentile charts to see how close they come to the "norm" or where they fall in comparison to the rest of the population.

Using the example above to test other parameters that are a public health concern, such as arsenic, lead, and nitrates, the statistics can be examined in light of the drinking water standards set by the U.S. Environmental Protection Agency (EPA). These standards are referred to as an MCL (maximum contaminant level), established for public water supplies. As seen in Table 2, none of the wells exceeded the MCL for those particular elements.

Visual aids such as pictures and graphs help people understand tables of numbers. For the visually inclined, the software offers a frequency histogram. Figure 1 depicts a frequency histogram for calcium values.

The correlations different parameters may have with each other can also be determined. An example is the correlation between magnesium and water hardness (mentioned earlier). Table 3 shows the results of a Spearman Rank Correlation result for these two parameters. Results indicate a significant correlation

Table 2. Groundwater Chemistry Analysis — Percentiles of 36 Parameters for SW Michigan Groundwater Project Wells

County: INGHAM
Township(s): ALL
Lithology or Aquifer: ALL

			Percentiles									
Parameter	N	Min	1	5	10	25	50	75	90	95	99	Max
Alkalinity	312	158.0	176.0	196.0	217.0	251.0	286.0	316.0	353.0	384.0	432.0	596.0
Aluminum	312	ND	ND	ND	ND	ND	ND	ND	ND	ND	0.08	0.52
Ammonia	311	ND	ND	0.03	0.04	0.07	0.14	0.21	0.33	0.46	0.76	1.09
Antimony	312	ND	ND	ND	ND	ND	ND	ND	ND	ND	ND	ND
Arsenic	293	ND	ND	ND	ND	ND	ND	0.007	0.013	0.018	0.025	0.042
Barium	312	ND	ND	0.010	0.030	0.070	0.100	0.200	0.400	0.400	0.600	1.400
Boron	312	ND	ND	ND	ND	ND	0.07	0.60	1.80	2.60	3.60	4.70
Cadmium	312	ND	ND	ND	ND	ND	ND	ND	ND	ND	ND	ND
Calcium	312	ND	1.2	8.9	33.0	62.0	76.0	91.0	122.5	156.0	289.0	339.0
Chloride	311	ND	ND	ND	ND	ND	ND	4.0	28.0	52.0	113.0	187.0
Cobalt	312	ND	ND	ND	ND	ND	ND	ND	ND	ND	ND	0.02
Chromium	312	ND	ND	ND	ND	ND	ND	ND	ND	ND	ND	ND
Conductivity	312	319.0	366.0	394.0	426.0	526.0	606.0	703.0	883.0	1124.0	1473.0	1567.0
Copper	312	ND	ND	ND	ND	ND	ND	ND	0.010	0.019	0.120	0.200
Fluoride	303	ND	0.1	0.1	0.1	0.2	0.3	0.6	0.9	1.1	2.2	2.7
Hardness	306	ND	23.0	75.0	160.0	242.0	290.0	350.0	432.5	525.0	649.0	700.0
Iron	312	ND	ND	0.08	0.20	0.40	0.70	1.30	2.60	3.30	7.60	9.20
Iron.2	0	ND	ND	ND	ND	ND	ND	ND	ND	ND	ND	ND
Lead	311	ND	ND	ND	ND	ND	ND	0.002	0.004	0.006	0.009	0.013
Magnesium	312	ND	0.40	2.90	9.50	17.00	24.00	29.00	35.00	40.00	48.00	73.50
Manganese	312	ND	ND	ND	0.01	0.01	0.02	0.04	0.07	0.15	0.60	1.13
Mercury	312	ND	ND	ND	ND	ND	ND	ND	ND	ND	ND	ND
Molybdenum	312	ND	ND	ND	ND	ND	ND	ND	ND	ND	ND	0.01
Nickel	141	ND	ND	ND	ND	ND	ND	ND	ND	ND	ND	ND
Nitrate	312	ND	ND	ND	ND	ND	ND	ND	ND	0.3	0.6	1.3

Table 2 (Continued). Groundwater Chemistry Analysis — Percentiles of 36 Parameters for SW Michigan Groundwater Project Wells

			Percentiles									
Parameter	**N**	**Min**	**1**	**5**	**10**	**25**	**50**	**75**	**90**	**95**	**99**	**Max**
pH	301	6.10	6.34	6.57	6.66	6.76	6.97	7.37	8.05	8.30	8.41	8.70
Phosphorus	312	ND	ND	ND	ND	ND	ND	ND	ND	0.1	0.2	0.7
Potassium	312	0.49	0.63	0.80	1.00	1.60	3.05	5.40	7.80	8.70	13.00	16.00
Selenium	312	ND	ND	ND	ND	ND	ND	ND	ND	ND	ND	ND
Silica	312	ND	ND	2.90	3.70	4.90	6.60	9.00	17.00	21.00	42.00	48.00
Sodium	312	1.60	2.50	3.20	3.60	4.90	8.40	22.00	76.00	148.00	209.00	313.00
Sodium.2	0	ND	ND	ND	ND	ND	ND	ND	ND	ND	ND	ND
Sulfate	312	ND	ND	1.0	3.0	8.0	19.0	54.5	98.0	148.0	415.0	743.0
Detergent	0	ND	ND	ND	ND	ND	ND	ND	ND	ND	ND	ND
Thallium	312	ND	ND	ND	ND	ND	ND	ND	ND	ND	ND	ND
Zinc	312	ND	ND	ND	ND	0.010	0.040	0.100	0.246	0.400	1.000	2.900

Note: Units are ppm except for conductivity (MMhos) and pH. ND indicates nondetection. Percentiles exceeding maximum containment levels are underlined.

From the Ingham County Health Department, Michigan, created by Western Michigan University.

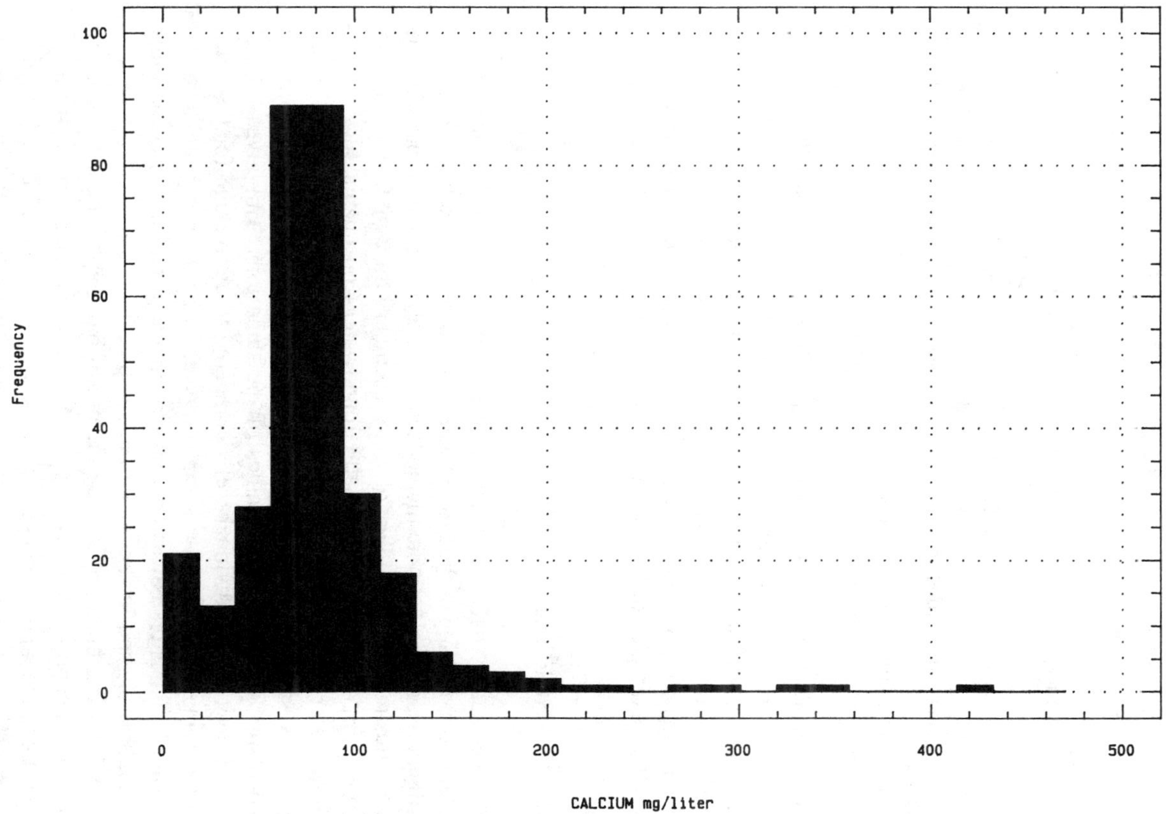

Figure 1. Frequency histograms created out of Statgraphics software for calcium values provide a graphic or visual analysis of data for quick review. The calcium data have a definite skewed shape to the left at approximately 80 mg/L. (From the Ingham County Health Department, Michigan.)

Table 3. Spearman Rank Correlations

	Magnesium	Hardness
Magnesium	1.0000	0.8560
	(312)	(312)
Hardness	0.8560	1.0000
	(312)	(312)
	0.0000	1.0000

Results show a significant correlation betwen well magnesium and hardness (using Statgraphics software
Coefficient (sample size) significance level

between the two in a direct trend: as one increases the other increases. To illustrate this trend, a simple scatter plot of the data (Figure 2) can be performed.

With the help of several statistical programs working on the chemical database, a very useful profile of groundwater quality can be made. How far the database user can legitimately go with the numbers will depend on his/her training in statistics. Other examples of the use of statistics are shown in Chapter 11.

3. GRAPHIC PROFILES

When water quality data are presented pictorially or graphically, one gains a much fuller understanding of the groundwater system. For a variety of groundwater investigations, several graphic methods are available for visual inspection or analysis. In illustrating the graphics of water chemistry the authors used another software program developed by the Hall Groundwater Consultants, Inc.[3] Figures 3 to 5 show water test results from different well locations in Ingham County, Michigan. These samples were compared to an "averaged" water well chemistry composed of the statistical mean values of the same ions. This gives a quick inspection of how these individual wells compared to an average for the entire county. The Bunkerhill Township sample appears to be the most similar to the average. The Williamston Township well indicates a difference based on much higher sodium levels and lower calcium and magnesium results.

Two common graphical profiles are the use of bar graphs and circular or pie graphs. A bar graph presentation is shown in Figure 4. For each sample, the major ion composition is in equivalents per cubic meter or milliequivalents per million. Figure 3 represents the same water test as a pie graph or diagram. The radius of each pie indicates concentration in parts per million.

Another method is the Stiff diagram, named after the hydrogeologist who first used it,[4] shown in Figure 5. This figure depicts five samples showing very different graphical patterns of major ion concentrations in milliequivalents per

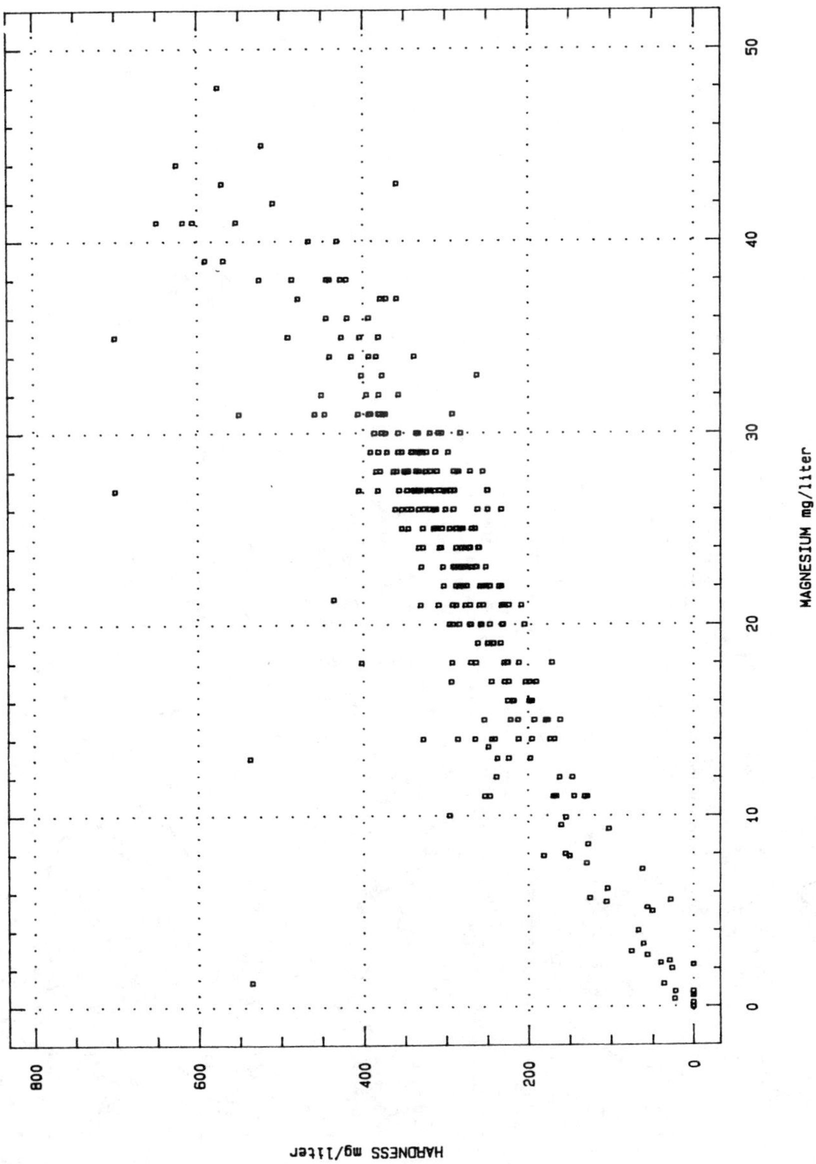

Figure 2. A scatter plot drawn with the Statgraphics software of water hardness to water magnesium demonstrating the same positive correlation between the two. With the exception of some data points, much of the data fall on a straight line. (From the Ingham County Health Department, Michigan.)

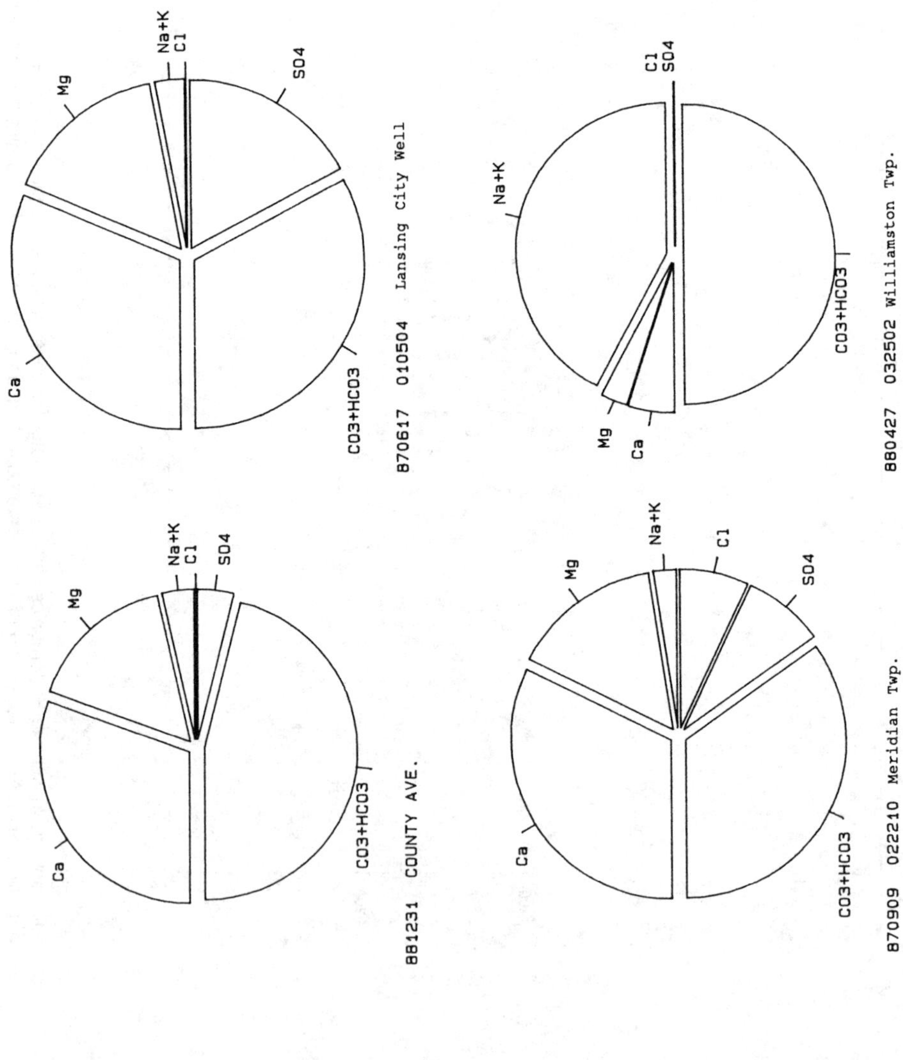
881231 COUNTY AVE.
Ca
Mg
Na+K
Cl
SO4
CO3+HCO3
870617 010504 Lansing City Well
Ca
Mg
Na+K
Cl
SO4
CO3+HCO3
870909 022210 Meridian Twp.
Ca
Mg
Na+K
Cl
SO4
CO3+HCO3
880427 032502 Williamston Twp.
Mg
Ca
Na+K
Cl
SO4
CO3+HCO3

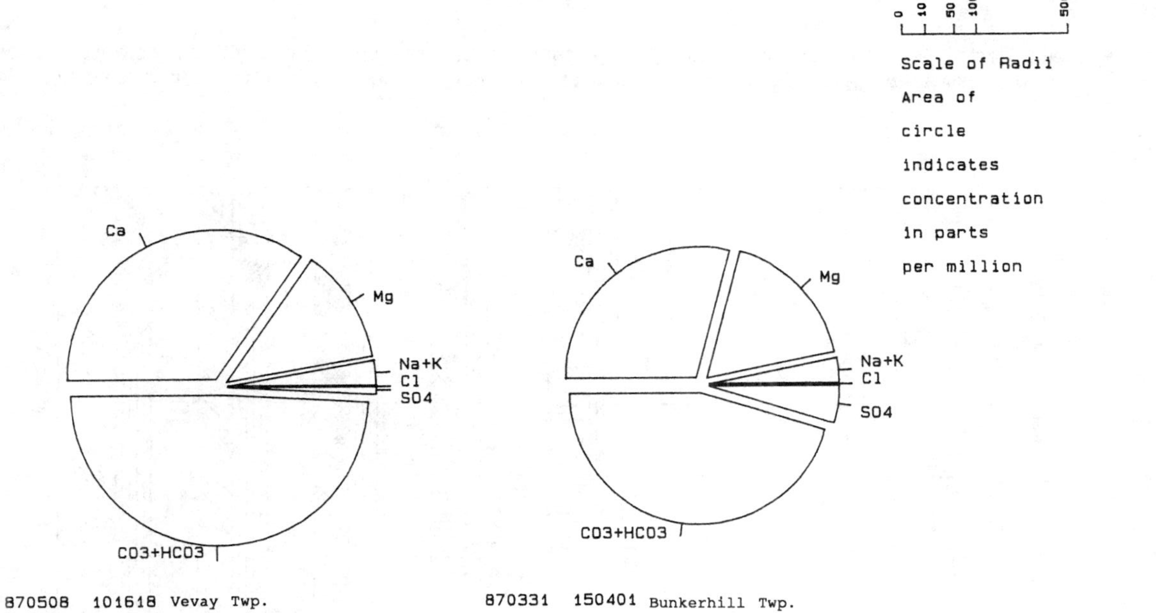

Figure 3. Using the same wells and softwater shown in Figure 3, pie graphs are made for comparison. A "County Average" data well is included for comparison. This "County Average" well is actually made up of average statistical data from the survey of Ingham County. (From the Ingham County Health Department, Michigan.)

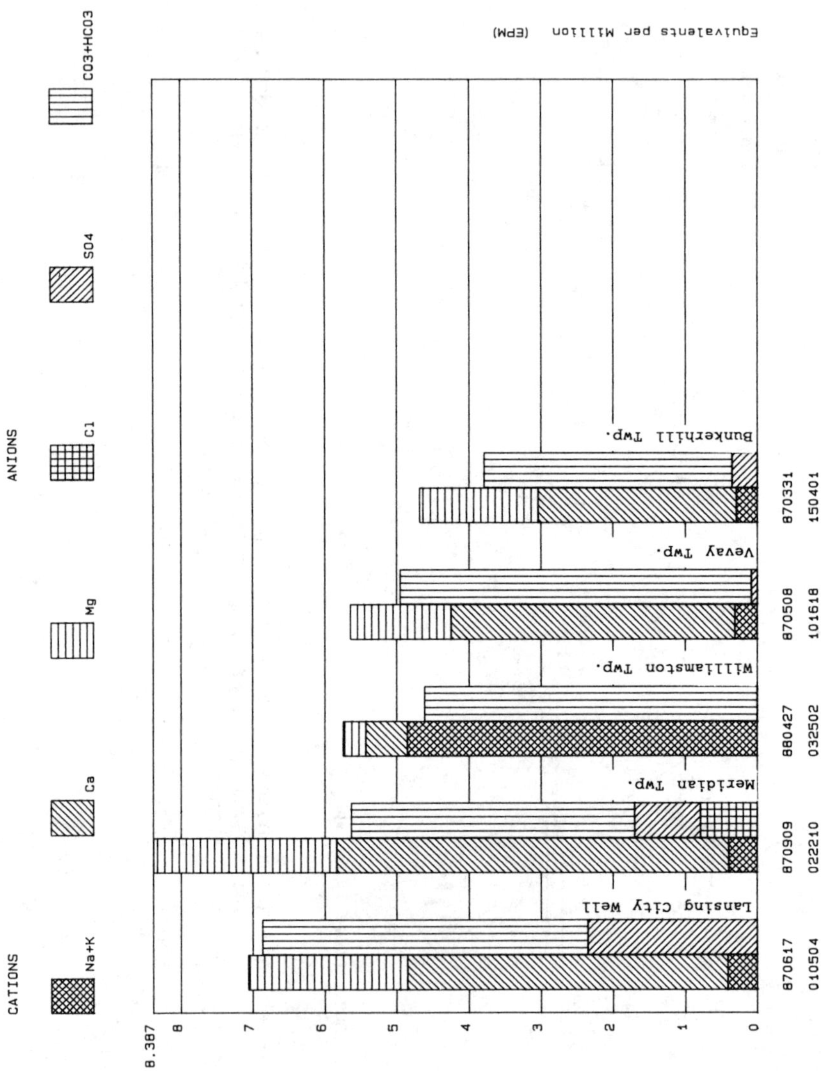

Figure 4. Water quality data taken from five wells in the Ingham County database are used with the Hall Groundwater Consultants' software to produce this bar graph using the major cations and anions in the water. (From the Ingham County Health Department, Michigan.)

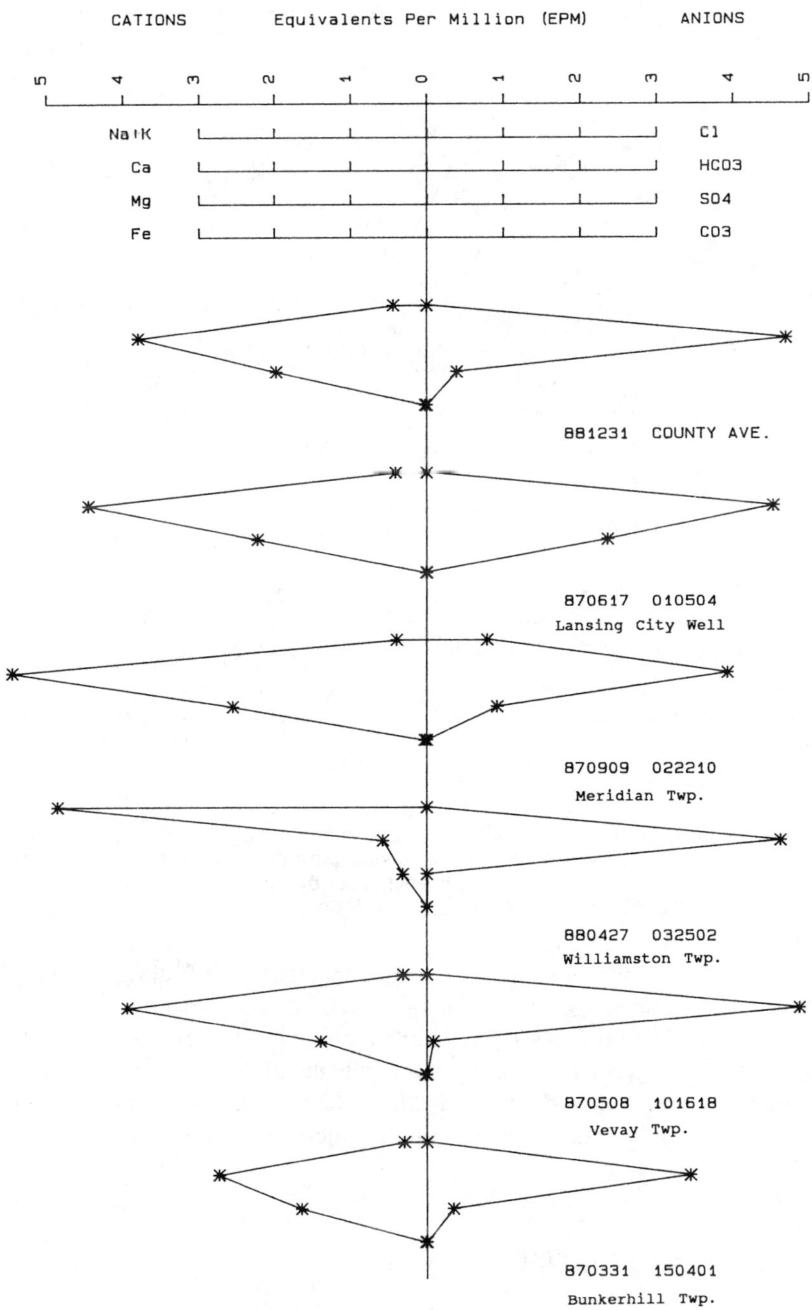

Figure 5. The five wells from Figure 4 are compared to the Ingham County average using STIFF diagrams made from the Hall software. (From the Ingham County Health Department, Michigan.)

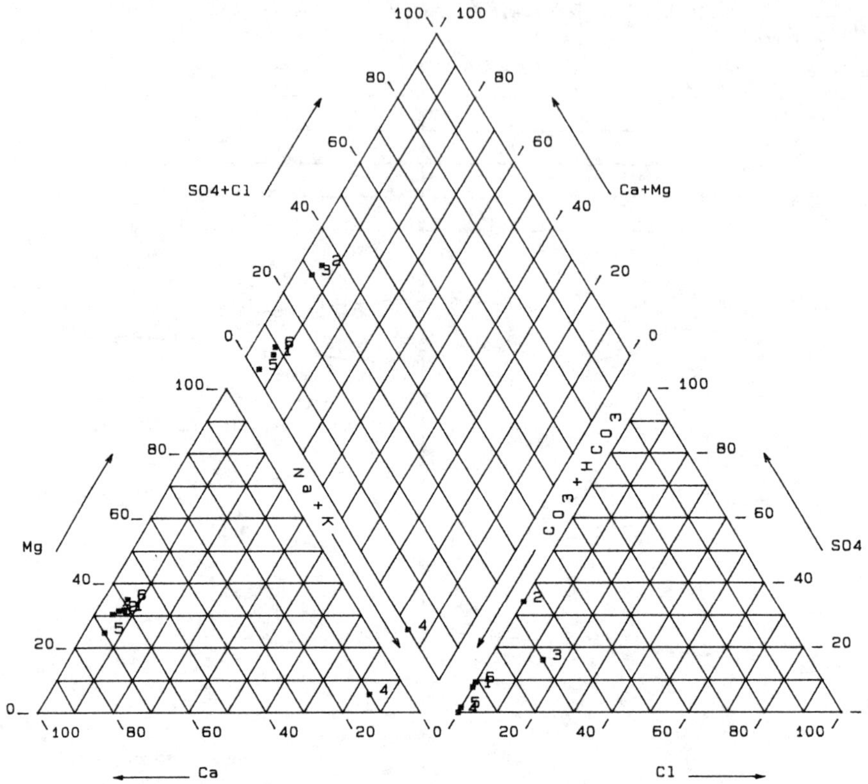

Figure 6. Piper diagrams are helpful for displaying many well chemistry samples on one graph. Chemical parameters or anions and cations for each well can be visually inspected as they compare to each other by chemistry on the graph. (From the Ingham County Health Department, Michigan.)

liter. Showing two or more water test results together as Stiff diagrams gives the observer a convenient and quick method for visual comparison.

To look at a large number of wells, the Piper diagram, developed by Hill and Piper,[4] is more convenient. Figure 6 represents the plotting of several wells onto the Piper diagram. The Piper[5] diagram allows the database user to observe several samples on a single graph so any possible patterns or trends in the data can be quickly observed.

4. GEOGRAPHIC DISTRIBUTION

A software program called C-Map,[6] developed at Michigan State University, is used in a geographic information system (GIS) application. C-Map uses a dBase file structure and allows for several files (or GIS layers) to be displayed geographi-

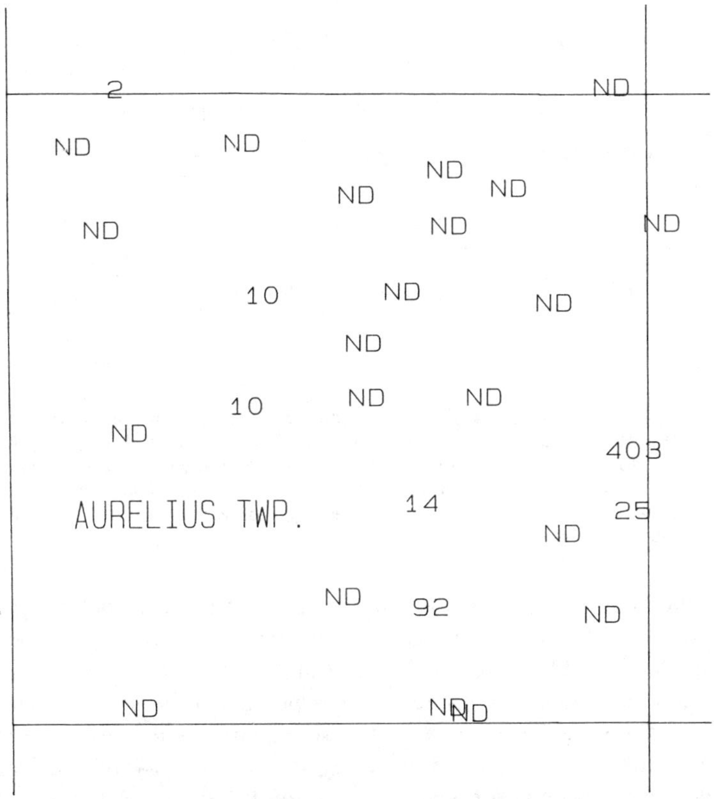

Figure 7. Using C-Map, a plot of the chloride levels in Aurelius Township was made. Most wells have values of 10 ppm or less. The high value of 403 indicates a possible pollution problem. (From the Ingham County Health Department, Michigan.)

cally. Figure 7 shows the distribution of chloride levels in Aurelius Township, Ingham County, Michigan. As expected from the statistics shown earlier, the results show what is normal for Ingham County — chloride levels less than the laboratory detection limit of 10.00 ppm (indicated as -10.0); however, the map does show a few wells that exceed this level. One well indicates a test level as high as 403.0 ppm. Because this represents a level far above the norm, the database user may want to investigate this area further and sample additional wells. Does this elevated level represent a natural condition, or is there a pollution problem? After completing a site investigation, it was determined that groundwater was polluted by the loss of high salinity water from an injection well used in conjunction with an oil well. At least two wells were found to be impacted by this salt. Figure 8 shows the distribution of chloride levels from other wells in the same area, including the two affected wells.

Another example that uses the information from the statistical analysis to map

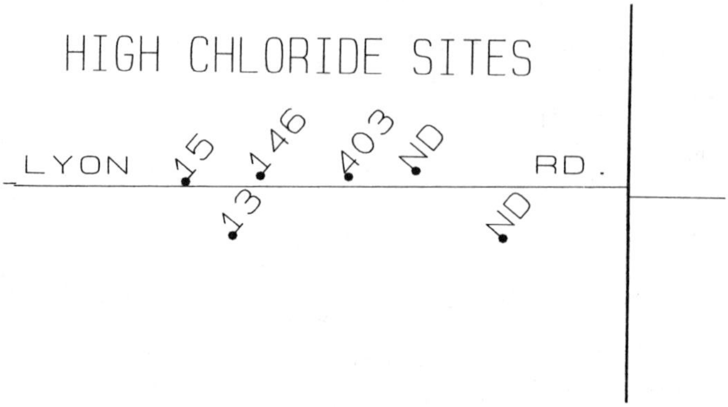

Figure 8. From the information shown in Figure 7, a further investigation of the area around the high chloride well (value of 403) was conducted. This plot shows the location and distribution of other chloride levels in wells nearby. It was determined that the two wells showing values of 146 and 403 ppm chloride were impacted by a spill of salt water at an injection well. (From the Ingham County Health Department, Michigan).

only those levels of one parameter that fall out of the range considered normal for an area. Using the Ingham County database, the range for any one parameter, for example, hardness, can be reviewed. Looking at an average level for hardness one obtains a value of 295.8 (see Table 1). Data points can be selected out of the database based any limit set for the data values. For this example, the authors examine what might be considered "naturally soft wells," or wells considered far below that which is normal for water hardness in the county. Based on a review of the statistical data, the database userchose a cut-off value of 200.0 ppm hardness as $CaCo_3$. Well data from the chemical database are selected based on this limit and then mapped using C-Map (Figure 9). By visual observation, a pattern for the "low hardness wells" can be seen.

Observing a pattern for the "soft" water wells the database user can look more closely into these particular areas to see what makes them unique. Most of the sampled wells in this area have very low hardness levels, <100 ppm as $CaCO_3$. Looking at the well-log database, the predominant bedrock type is shale. Could shale bedrock acting like a water-softening device be exchanging sodium for calcium and magnesium in the aquifer system? Using the mapping program can lead to further discoveries, but also more questions.

Using a mapping program such as C-Map can give quick visual access to the data for inspection. More examples of this application and the use of statistics and graphics are covered in Chapter 11.

Once a profile of the water chemistry is made, site investigations and future research into the groundwater of an area can be enhanced by the rapid retrieval of known hydrogeologic information. The database can provide preliminary information to help direct the researcher or investigator. It can also provide a comparison against which new data can be reviewed.

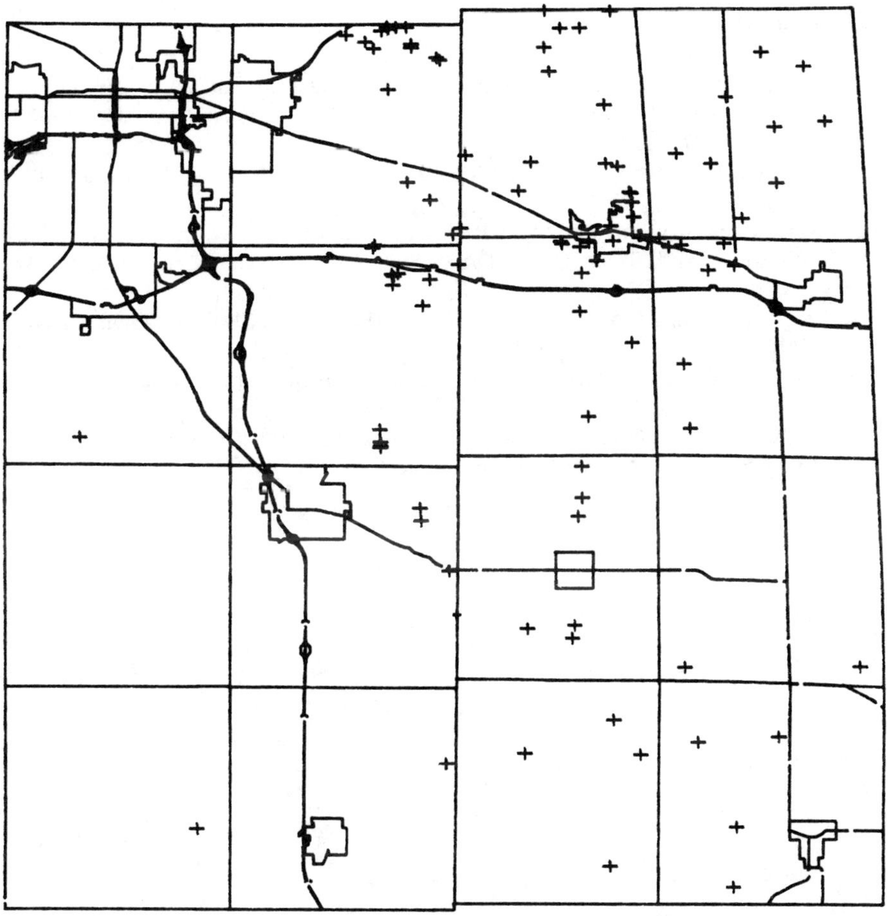

+ Well Site

Scale = 1: 250000

Figure 9. A map of Ingham County showing the location of "naturally softened" wells (<200 ppm hardness as $CaCO_3$). Most of the wells are located in the upper northeast quadrant of the county with a significant number in the lower southeast quadrant. (From the Ingham County Health Department, Michigan.)

REFERENCES

1. Geobase4, Cross-section software by Hall Groundwater Consultants, Inc., Earthware of California, 30100 Town Center Drive Suite 0196, Laguna Niguel, CA, 92677 (714)495-5727.
2. Statgraphics Version 3.0, STSC, Inc., 2115 East Jefferson St., Rockville, MD, 20852.
3. Water Chemistry Software by Hall Groundwater Consultants, Inc., Earthware of California, 30100 Town Center Drive, Suite 0196, Laguna Niguel, CA, 92677 (714)495-5727.
4. Freeze, R. A., and J. A. Cherry. Groundwater (Englewood Cliffs, NJ: Prentice-Hall, 1979).
5. Piper, A. M. "A Graphics Procedure in the Geochemical Interpretation of Water Analyses," *Trans. Am. Geophys. Union* 25:914-923 ( ).
6. C-Map Computer Mapping, Display and Analysis, User Manuals version 1.03, Center for Remote Sensing, Michigan State Univesity, East Lansing (1989).

CHAPTER 10

Management of the Database – Electronic Libraries

1. THE DATA/MAP/GRAPHICS RESOURCE SYSTEM

Once electronic files are installed,an organization quickly finds they are reconciled to them and never return to paper systems; however, maintenance routines and new habits are required that may not be part of any employee's job description. A manager should assign the job of Data Librarian and allocate time for that person to document and maintain the files.

Databases acquire value with use, and to use them means to keep them well documented. The main job as librarian is to organize a data/map/graphics resource system. A county resource system should contain:

- •Data — Floppy disk(s) or other data media labeled and cataloged. Data categories include: map layers (lakes, rivers, roads, etc.), well construction, chemistry databases, and plot files used to make particular illustrations.
- •Maps — An atlas can cross-reference these paper maps with the electronic files used to create them. An index provides the locations of master copies of the maps.
- •Graphics — Screen graphics should be captured as PIC or PCX files so they can be used for slides, presentations, animation, or desktop publishing. Prepare procedures to instruct people in how to "capture" the images they have made and add them to the screen graphics library.

Once the labor has been invested in building the databases and taking the water samples, it is advisable to protect the data by storing a complete set at another location, in case the building in which it is housed is damaged. Computers can malfunction, and backup diskettes are lifesavers.

2. THE GEOGRAPHIC INFORMATION SYSTEM

The libraries of data can serve as a geographic information system (GIS). In a GIS, a well construction database would be considered as a data layer, as would the chemistry results database. Related layers (or databases) can be acquired to help with planning, environmental response, economic development, emergency response, land use, transportation, etc. GIS technology will affect an entire spectrum of professional disciplines in the future.

To add GIS strength to the groundwater databases, a company should acquire the following layers (or databases):

Superfund pollution sites
Soils
Septic tank land application sites
Road salt storage areas
Railroad and abandoned railroad beds
Landfills
Municipal and irrigation wells
Lakes, rivers, streams and wetlands
Land use
Nitrogen application per acre
Roads and highways
Political boundaries

Figure 1 shows a fishnet map made from the well construction database (fields: static water level and elevation), the municipal boundaries database (city names), and the databases for rivers and streams. The well data comes from Calhoun County's participation in the Michigan Groundwater Survey,[1] and the political labels, rivers, and streams layers from the Michigan Resource Inventory Program (MIRIS).[2]

The potential of GIS is great, but the process of building one is demanding. New skills must be learned and databases must be built, bought, or acquired. There is a wide variety of GIS software, and a survey comparing them was recently published by GIS World. To help develop the kinds of people who will become skilled GIS operators, the authors have reproduced that survey as well as a list of common GIS functions in Appendix E.[3] A diagram[4] showing the overview of database and map work involved in a GIS is shown in Figure 2.

The 1990 census will open a new world of data and digital maps to the GIS user. Now and in the future, the challenge will be to match databases and GIS tools. Because the GIS field is new, at the present time much effort is being spent on moving data from one format to another. Figure 3 shows Michigan's working diagram of GIS data formats and translation routines.[5]

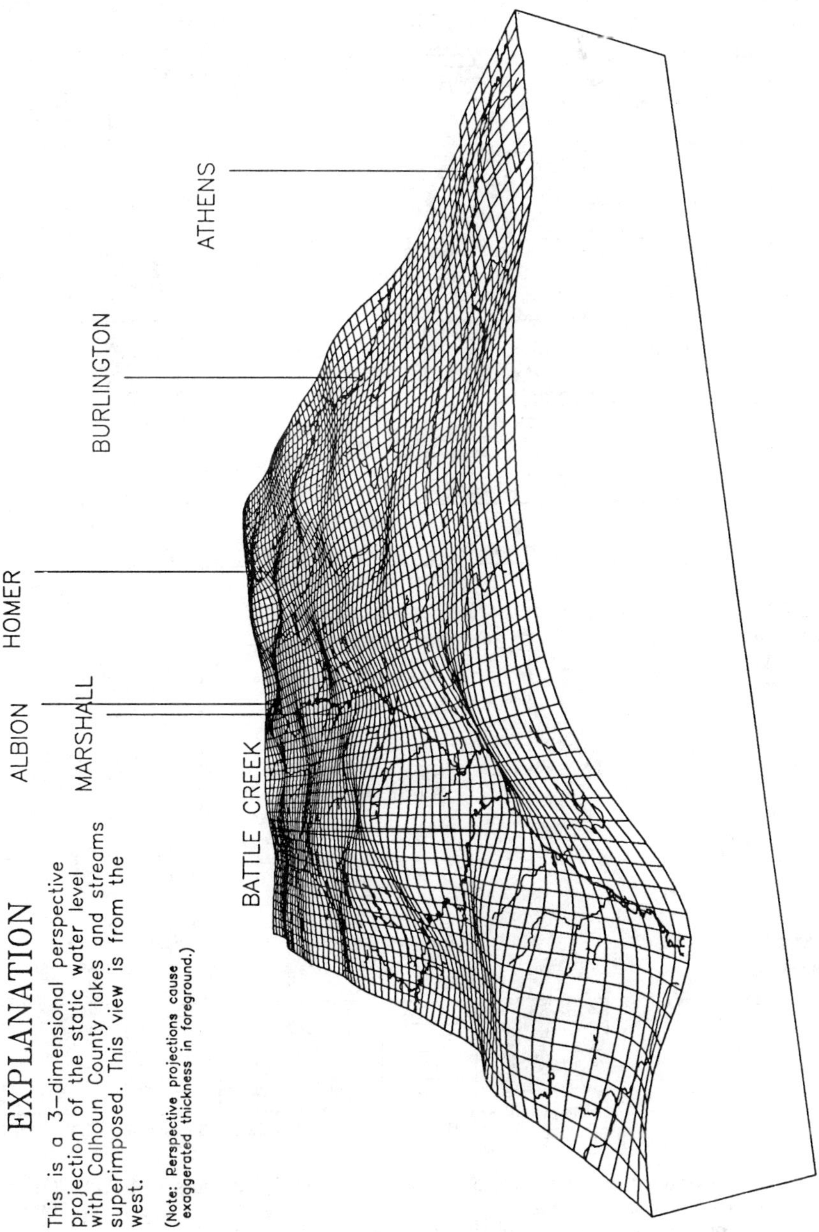

Figure 1. Several database layers are combined to make one map. Static water level data came from the Michigan Groundwater Survey, while municipal locations and rivers and streams came from Michigan Resource Inventory Program. (From the Institute of Water Sciences, Western Michigan University, Kalamazoo.)

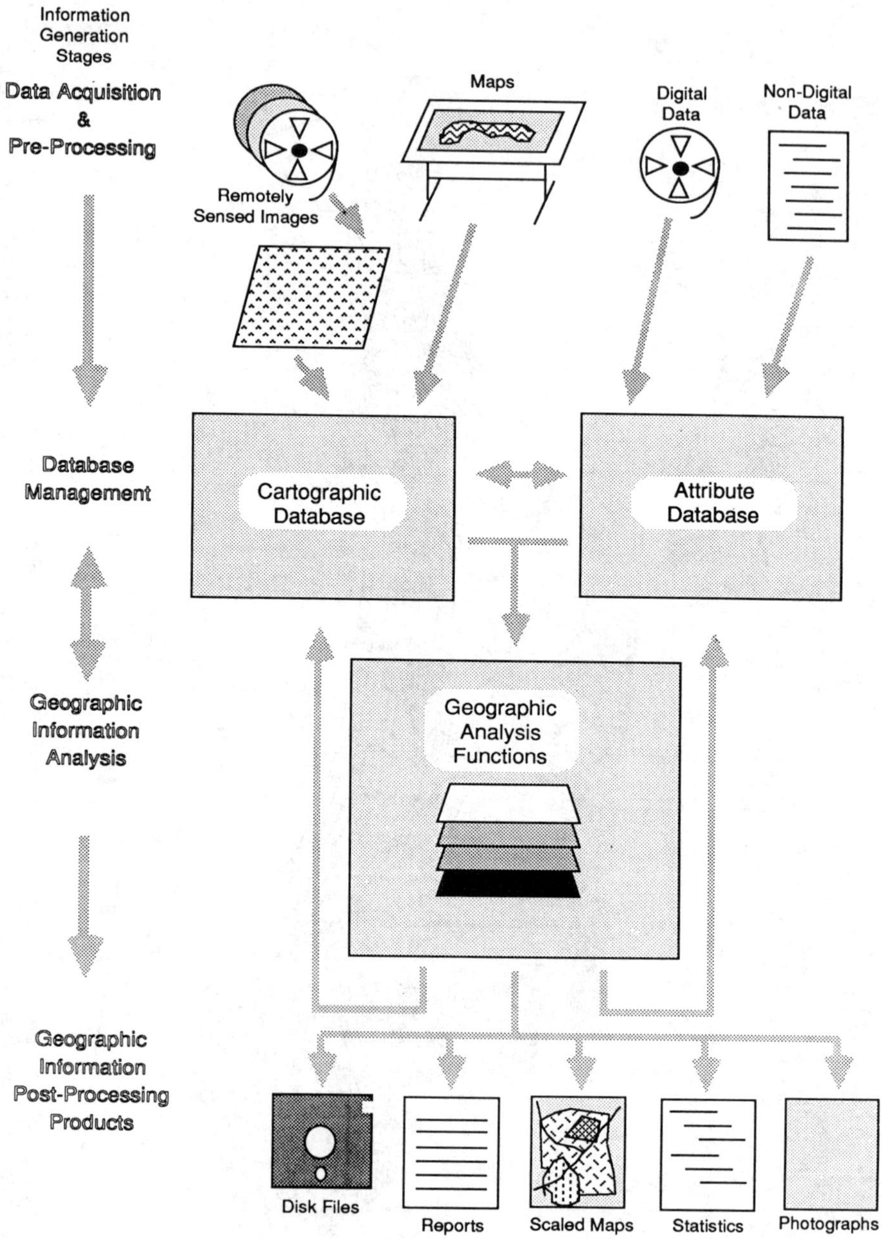

Figure 2. Overview of information processing and expected outputs of a GIS. (From GISThemes: Newsletter of the GIS Pilot Projects Module by G.E.M. Center for Groundwater Education at Western Michigan University.)

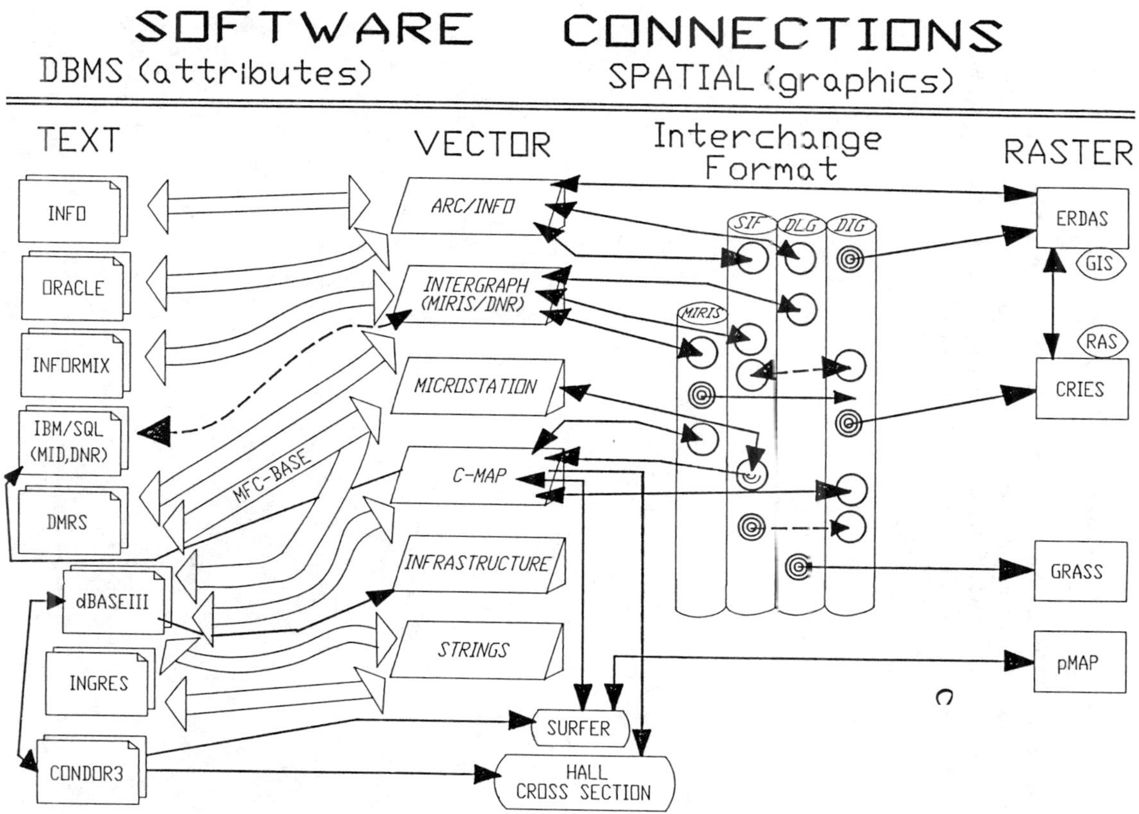

Figure 3. The complexity of data exchanges between software packages. This shows Michigan's working diagram of an ever-changing set of relationships. (From the Center for Remote Sensing, Michigan State University, East Lansing.)

REFERENCES

1. Mosher, J., and R. Passero. *Geology of Calhoun County* (Kalamazoo: Western Michigan University, 1990).
2. Michigan Resource Inventory System (MIRIS). Department of Natural Resources, State of Michigan, Lansing.
3. *The GIS Sourcebook,* GIS World, Inc., P.O. Box 8090, Fort Collins, CO 80526, 1989, 1-800-GIS-WORLD.
4. Dickason, D. GISThemes: Newsletter of the GIS Pilot Projects Module by G.E.M. Center for Groundwater Education at Western Michigan University, 1:7 (1989).
5. Enslin, W., and D. Lusch. Software Connections, Center for Remote Sensing, Michigan State University, East Lansing (December 1988).

CHAPTER 11

Impact – Practical Applications

A tool is only good as the person(s) using it. A trained carpenter can build a shed to store machinery and building materials. A crew of trained carpenters can build a house for a family to live in. A crew of properly trained staff can create an effective groundwater management system that can protect the water supply of a community.

This chapter presents several examples of the ways in which a groundwater database can be used effectively to both protect the groundwater and the environment and assist in the investigation of a suspected pollution problem. The goal of this chapter is to provide education, motivation, and perhaps inspiration to the potential users of a groundwater database. The justification for a database will come with its effective use in protecting the environment.

Chapter 7 mentioned the difference between the terms "contamination" and "pollution." Pure water, the presence of H_2O only, is generally not found in nature. Water will always be "contaminated" by something. In surface water or groundwater the water quality is associated with a number of dissolved minerals, bacteria, sediment, etc. When the term "contamination" is used it refers to the concept of something other than H_2O being present. On the other hand, when the term "polluted" is used it refers to the condition in which the water has been rendered unusable for drinking, detrimental to animals and the environment, or put in an unusable condition for any specific use.

All of the examples discussed come from case studies in Ingham County, Michigan, as the authors have done most of their practical work in this area. Because of differences in geographic location and geology, these examples may not specifically apply to some other areas of the U.S., although the basic concepts should.

1. THE PROPOSED INDUSTRIAL PARK

The development of land for industrial facilities is an ongoing process in the U.S. In the past, little work was actually done to carefully examine the potential dangers to the environment a proposed industry may pose. A larger concern was the availability of transportation routes, labor, taxes, land prices, etc. Because of the growing concern over the effect of industry on the environment, many states have passed legislation requiring existing and future industry to meet certain criteria in building industrial facilities. Criteria include secondary containment for chemical storage, the manifesting of certain wastes, and the development of programs to address a chemical accident.

The groundwater database can provide preliminary hydrologic information on a proposed site to those conducting the review of the site. Some preliminary questions that should be asked about a site include:

1. Does the site fall within or is it near a possible groundwater recharge area in which any possible spills or leaks of a chemical material would easily infiltrate to any groundwater aquifers?
2. Are there water wells in the area that could be adversely affected by a spill incident at the site?
3. Does the geology provide any protective layers to groundwater aquifers such as clay glacial drift or shale bedrock?
4. What is the current status of groundwater quality?

Figure 1 shows the location of a proposed industrial park site in Leroy Township., Ingham County. C-Map[1] was used to chart the location of several wells. Data from these wells were then applied to a graphic on cross-section software.[2] Figure 2 shows a geologic north-south cross-section using wells 81008, 81004, 81005, and 81502. The cross-section indicates a considerable amount of clay covering much of the proposed site, with some sand and clay mixed soils on the north and south site boundaries. The glacial drift material sits on top of a sandstone bedrock aquifer that most wells in the area use for a water source. No protective impermeable layers other than the clay layer help protect the sandstone aquifer. The glacial drift material offers some protection to the sandstone aquifer by virtue of its depth, which is approximately 75 ft. The cross-section also shows a sand-gravel well that may provide information on the water quality of the drift aquifers.

The static water level measurements from well construction records in the area of the proposed industrial park were gathered from the database and used with Surfer,[3] a contouring and mapping software package. A static water contour map for the park is shown in Figure 3, indicating that groundwater flow in the bedrock aquifer is generally to the NNE. Pollution that may occur at this park and then enter the bedrock aquifer may affect other water supplies in this direction.

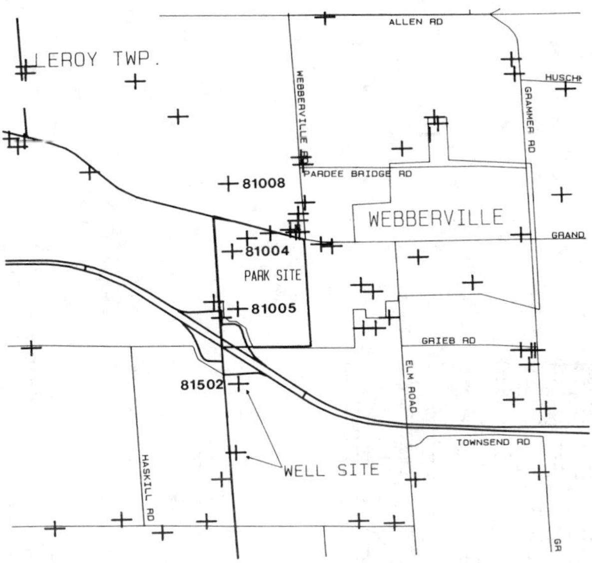

Figure 1. This site location map of a proposed industrial park indicates the location of several wells in the area. The four numbered wells will be used to plot a cross-section through the proposed site for review. (From the Ingham County Health Department, Michigan.)

A preliminary review of this information revealed that the proposed site offers some protection for groundwater because of the subsurface clay material found on the geologic cross-section. Facilities that use chemical products or generate chemical wastes would still be required to provide protection against accidental spills or releases, but this site offers more natural protection than a proposed site that may be composed of more permeable sandy subsurface material. A common geographic information system (GIS) function is to display map layers together. Figure 4 shows a C-Map plot of the site with soil data. The site can now be reviewed for the suitability of on-site sanitary waste systems (septic tanks and drainfields).

The surface soil map indicates that a majority of the site consists of Aubbeenaubbee-Capac sandy loam, Colwood-Brookston loams, Kibbie loam, and Marlette fine sandy loam soils. Most of these soils are severely limited in both on-site sanitary waste systems and building site development due to wetness, slow permeability, and low building strength. Because of these limitations it was recommended that the sanitary sewer system from a nearby municipality be made available to the park, and the developer would be required to provide engineering techniques to overcome the soil limitations for supporting a building.

Because a municipal sewer system was recommended it was also felt that municipal water be extended at the same time; however, if cost or other circum-

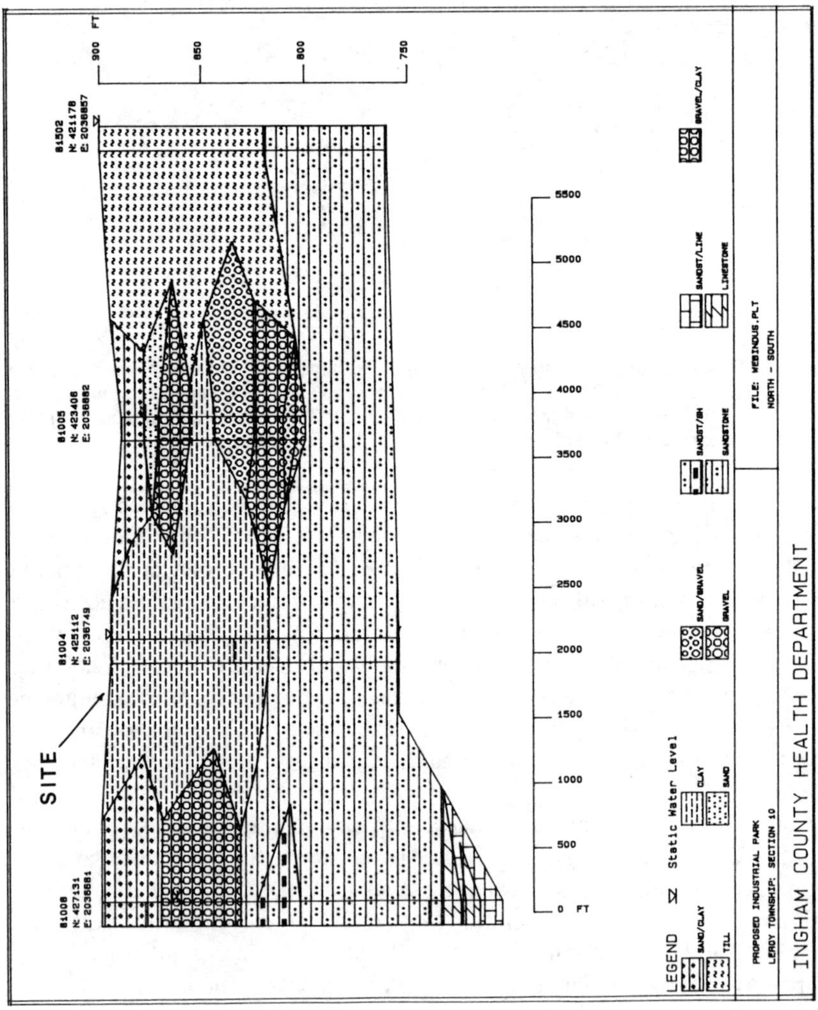

Figure 2. Cross-section from wells shown in Figure 1. The geologic information provided on the cross-section helps determine aquifer depth sites best suitable for building facilities, groundwater protection, aquifer depth, etc. (From the Ingham County Health Department, Michigan.)

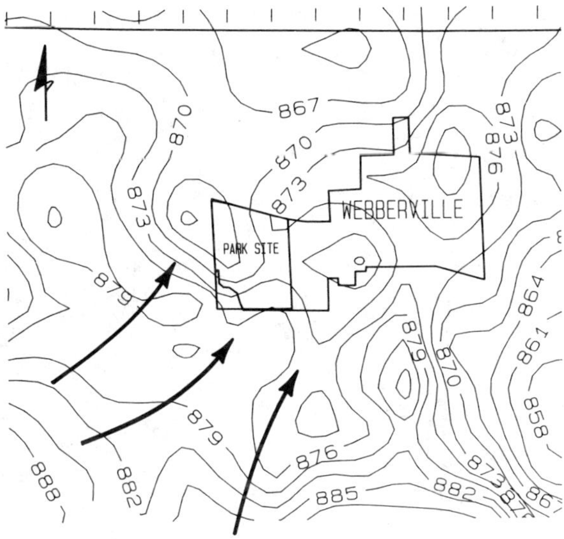

Figure 3. This static water contour map provides a general view of groundwater flow in the bedrock aquifer which appears to be to the NNE. In the event that a major spill of a polluting material finds its way to the bedrock aquifer, possessing this information will be important in remediation work. (From the Ingham County Health Department, Michigan.)

stances did not make this a possibility, then considerations for future on-site water supply wells must be made. To protect the well and groundwater, the wells should be grouted properly and the well casing set at a level or in a geologic strata that could isolate upper shallow aquifers from the lower aquifer used for the water source. The well construction records of current wells along with the geologic cross-section can aid well drillers and planners in determining the depth of the new wells, and suitable well construction. In such a case, new wells drilled into the lower bedrock aquifer are required, as well as extending the well casing into the bedrock and grouting it with neat cement or a high density bentonite material from the bottom of the well casing to surface. Isolation distances from any potential pollution source must also be addressed for well locations.

The water chemistry database provides a "background" file for water quality for this area prior to establishment of the industrial park, and provides information later if in future the water quality is altered because of the industrial park. Table 1 and Figure 5 show wells in the area that reflect common levels of water quality for Leroy Township, A closer inspection of the water test results indicates a slight difference between the one glacial drift well and the deeper bedrock wells (Table 1). The drift well show a measurable level of chloride but the bedrock wells do not. Other parameters such as sulfate, calcium, and magnesium are also slightly higher. This difference may be due to a natural condition or a pollution source such as road salting, and may suggest that any drift wells are more susceptible to surface water

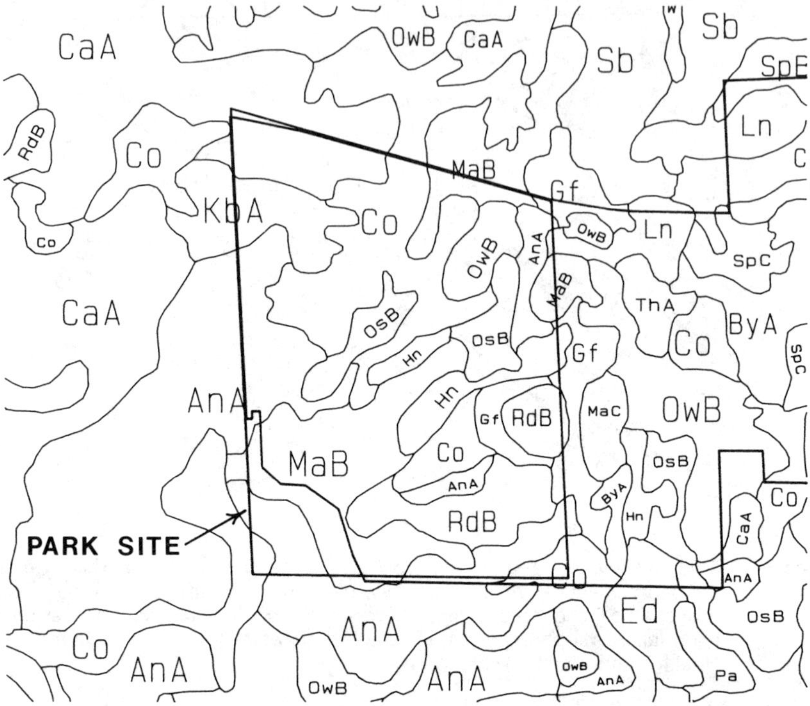

Figure 4. This soils type map for the proposed industrial park can be referenced to engineering data provided by the U.S. Soil Conservation Service for sanitary facilities, road construction, building foundations, ponds, etc. (From the Ingham County Health Department, Michigan.)

pollution. When building facilities for holding chemical products and hazardous wastes, developers will need to take this information into account.

In review, the groundwater database can address some preliminary concerns. The clay layers found at the site do offer some protection to the site. Sanitation for the site must be included in the nearby sanitary sewer system because of the limitations for on-site septic tank and drainfield systems. Bedrock wells do not indicate problems with water quality; however, shallow wells suggest a potential problem with a vulnerable glacial drift aquifer. By using such information, a municipal planning board can make better recommendations on such a site and establish a course of action. It must be emphasized that this kind of preliminary work is not a substitute for detailed on-site work, but should help direct the planning of further studies. The recommendation of any site review group should include on-site hydrogeologic research along with the preliminary information gathered from the groundwater database.

Table 1. Proposed Industrial Park Wellwater Results

Well site	Ca	Mg	Na	K	ALK	SO_4	Cl	NO_3	Fe
081003	73.9	25.5	12.0	4.0	296.0	7.0	0.0	0.0	0.6
081004	76.6	23.3	14.0	4.8	298.0	7.0	0.0	0.0	0.4
081005[a]	105.2	36.8	6.0	1.6	309.0	61.0	12.0	0.0	1.3
081007	70.0	21.0	14.0	4.3	288.0	16.0	0.0	0.0	2.4
081008	74.7	29.0	8.0	2.5	302.0	8.0	0.0	0.0	0.9
081009	75.4	27.9	10.0	3.5	304.0	9.0	0.0	0.0	1.5
Cnty ave.	79.7	23.6	25.5	3.4	287.0	45.2	12.7	0.0	1.1
Twp. ave.	63.0	21.4	35.2	4.2	305.6	19.1	5.6	0.0	0.9

Note: Water quality results collected for the proposed industrial project will provide "baseline" data on the park area for future comparative reference. It also provides data to those facilities that require certain water quality criteria in the facility. These data are compared to average values found for both the county and the township. Significant differences were noted for the one glacial drift well (ID# 081005) in comparison to the bedrock wells. Some smaller differences were also noted between the bedrock wells and the county and township average.

[a] Glacial drift well.

From the Ingham County Health Department, Michigan.

2. THE PERMITTING OF A WELL IN A SENSITIVE AREA

Both local and state agencies throughout the U.S. conduct programs that regulate the installation of water supply wells through a permit process. With the increasing awareness of groundwater pollution problems, many groups or agencies attempt to assess known, suspected, or potential sources of groundwater pollution before a well is installed near a source. If the agency has access to some basic hydrogeologic information, stipulations on the permit for a well regarding location, construction, and water testing can be included.

Figure 6 shows the location of a proposed public water supply for a new commercial facility near a municipal wastewater lagoon site in Dansville, MI. The local health agency raised concerns about the site because of the possibility of leachate moving from the lagoon site to the proposed well. Cross-sections from the records of several wells in the area are shown in Figure 7. Glacial drift material is made of a mixture of sand, clay, and gravel that lies atop a shale bedrock material. Most of the wells in the area appear to access a lower sandstone bedrock aquifer for water. The shale bedrock layer on top provides the best protection for this source of water. Groundwater chemistry for wells currently in the area does not indicate unusual water quality (Figure 8). The static water level data from wells in the general area provide a static water contour map for the bedrock aquifer (Figure 9). General groundwater flow in the bedrock is to the NNE, fortunately

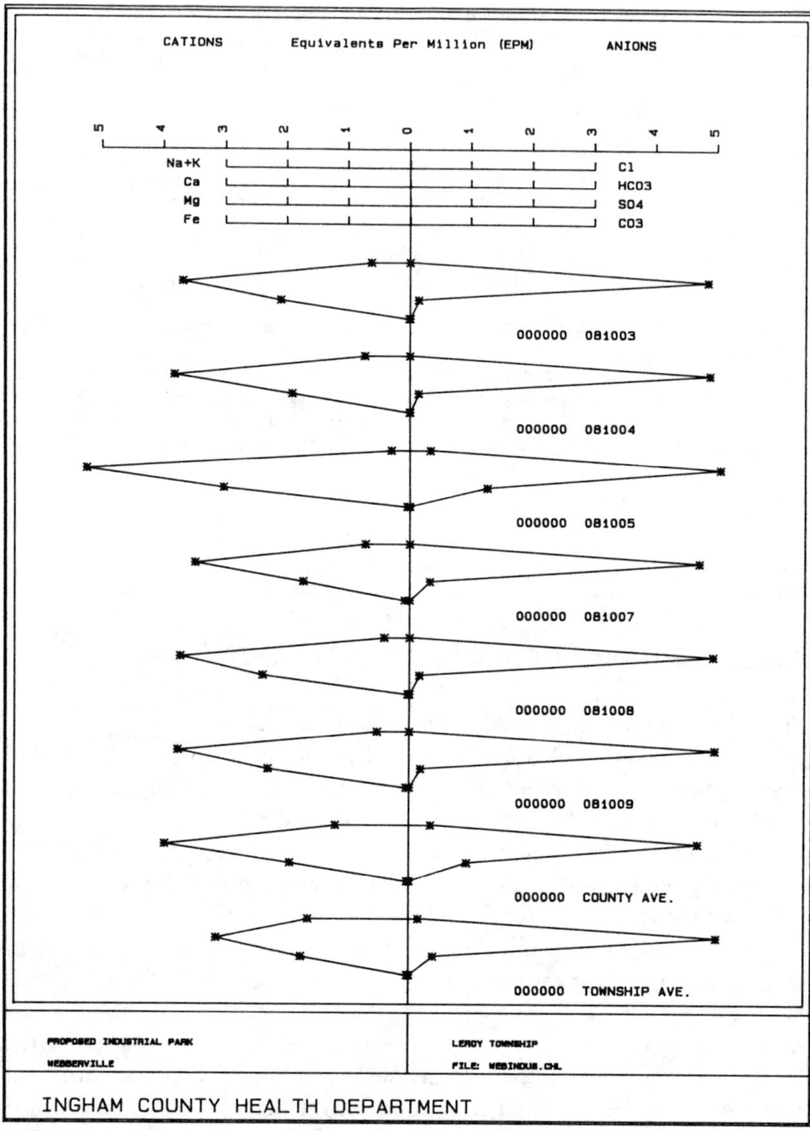

Figure 5. The same six wells from Table 1 shown as STIFF diagrams in comparison to the county and township average. Differences were noted again between the glacial drift well (ID# 081005) and the bedrock wells and also between the site wells and the county and township averages. (From the Ingham County Health Department, Michigan.)

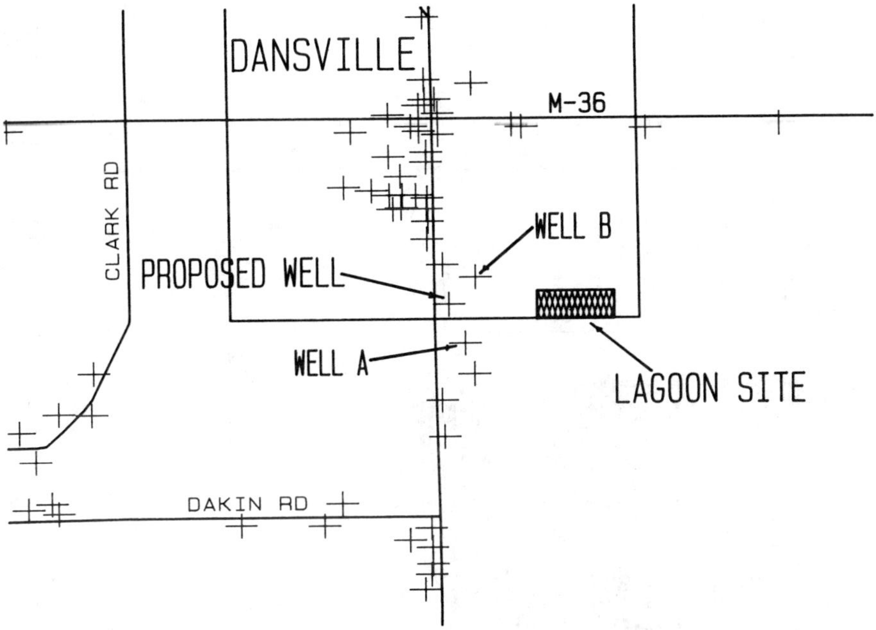

Figure 6. C-Map was used to show the positions of existing wells against the proposed well and the site of potential groundwater pollution, the sewage lagoon. Wells A and B will be included in a cross-section of the site lithology. (From the Ingham County Health Department, Michigan.)

away from the proposed well. The site map indicates a nearby creek to the ESE, which flows to the NE. Shallow aquifers may follow the surface topography to locations where surface water is found; therefore, it is likely that groundwater flow in the glacial drift aquifers follows the same pattern as the creek: away from the proposed well.

Using this information, a permit for the well was approved with the following stipulations:

1. The well is to be drilled into the lower bedrock aquifer with the well casing to be extended into the first shale bedrock layer indicated on the geologic cross-section.
2. The well will be grouted from the bottom of the casing to ground surface to help protect and isolate the upper more vulnerable aquifer from the bedrock aquifer.
3. The well is to be sampled twice a year to monitor any changes in water quality.
4. The well will be located near the building following local requirements for isolation distances.

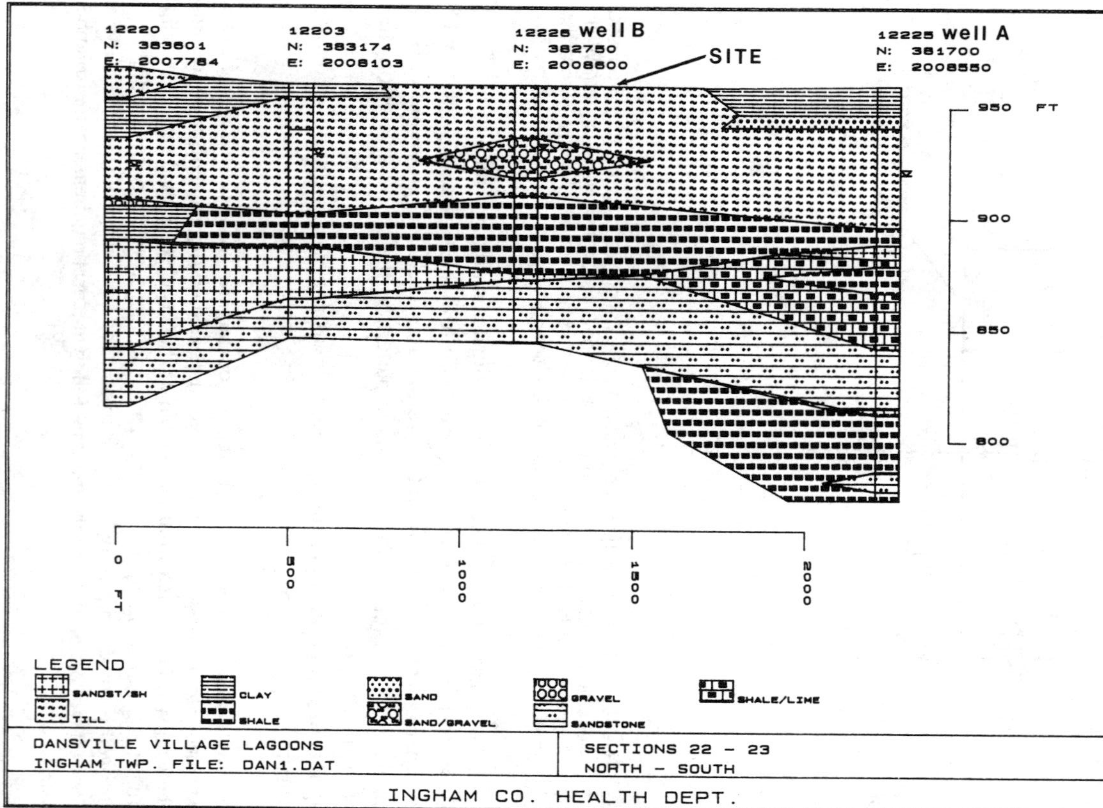

Figure 7. This cross-section, which includes wells A and B, indicates the lithology structure near the proposed well. This information is helpful in determining well location and construction that will provide some protection against pollution. (From the Ingham County Health Department, Michigan.)

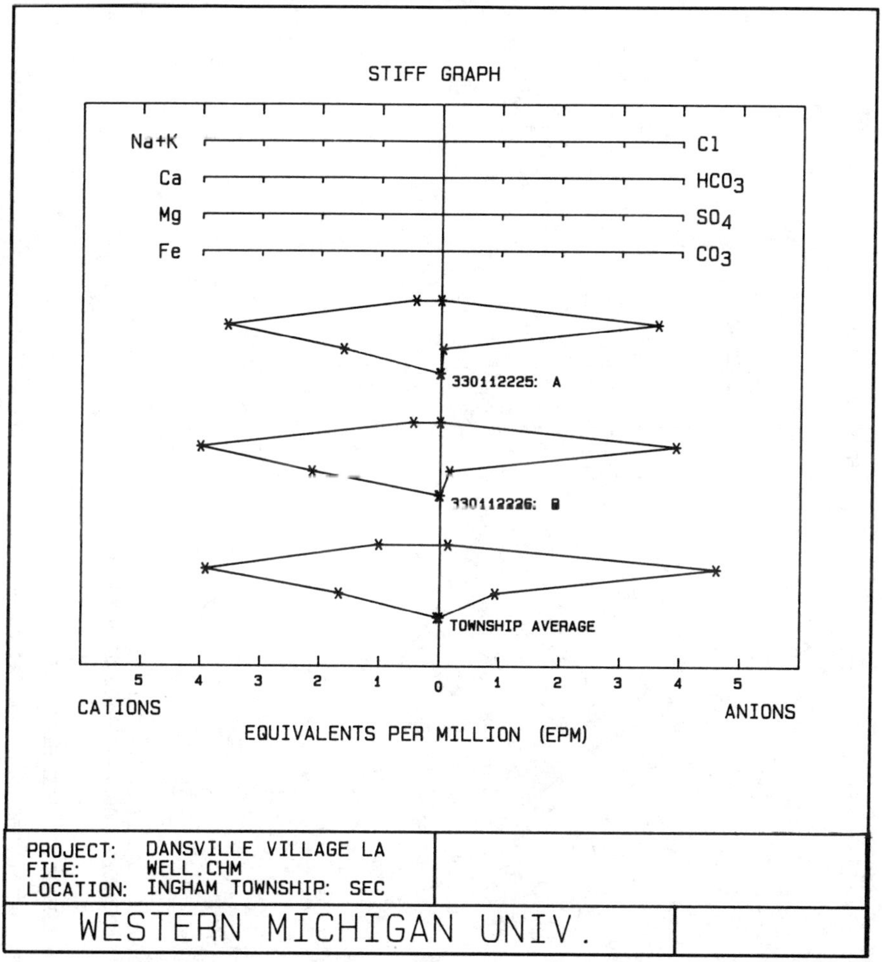

Figure 8. The water chemistry from wells A and B were compared to average values for the township. Groundwater appears to be somewhat different than "average" groundwater but did not indicate any pollution problems. (From the Ingham County Health Department, Michigan.)

These stipulations can be reviewed by both the well driller and facility owner prior to construction. This information should help prevent pollution problems, and the construction guidelines should provide additional protection from a potential pollution source.

3. SALT POLLUTION

Road salting in various parts of the U.S. has been a standard and effective

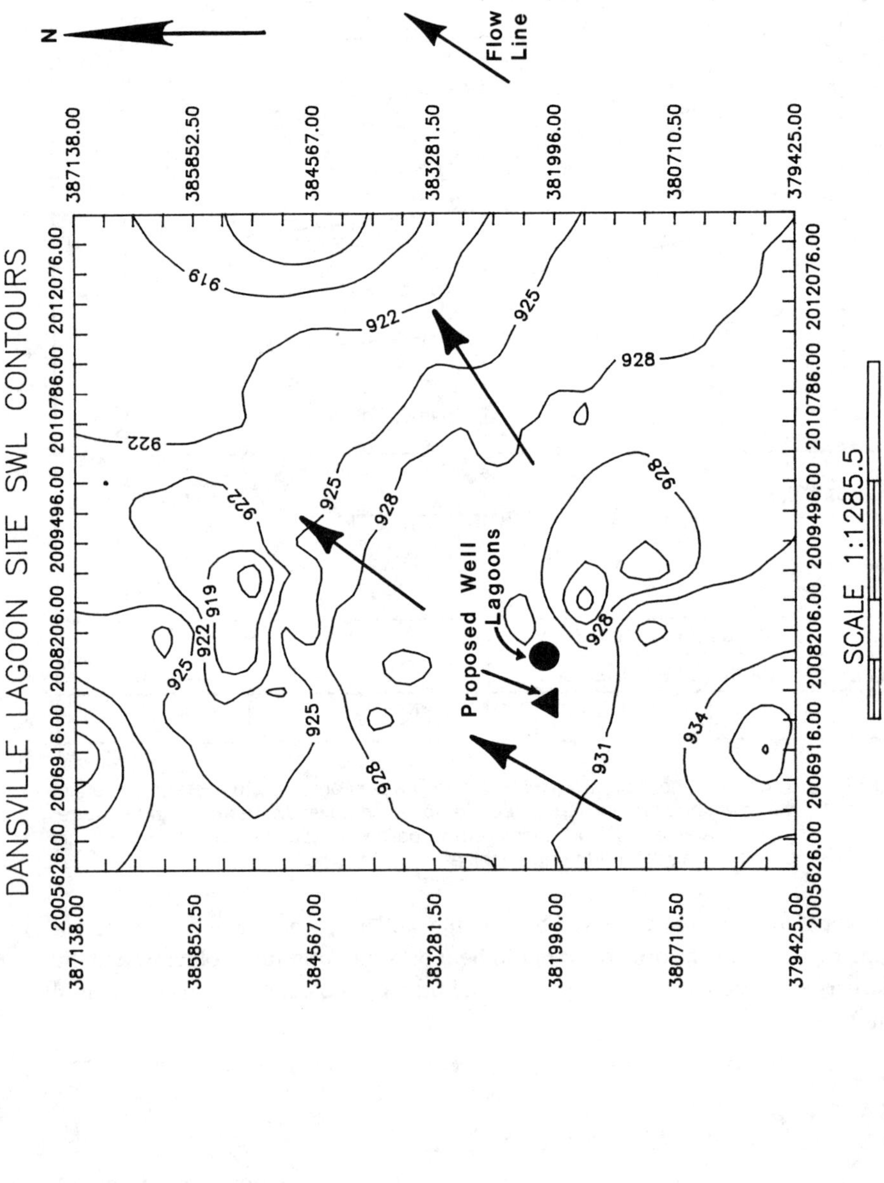

Figure 9. A static water level contour map drawn from Surfer, which indicates that general groundwater flow in the bedrock aquifer is to the north and northeast away from the proposed well. (From the Ingham County Health Department, Michigan.)

procedure for ice control on roads and highways for several years; however, several wells have become polluted as a direct result of this procedure.

Table 2 is a percentile table for several parameters in Ingham County. A majority of wells showed chloride levels <10.00 ppm. When the sampled wells were mapped, several in the vicinity of the City of Mason had very high chloride levels. Further investigation led to the sampling of more wells which indicated a possible salt pollution problem. It was discovered during the course of the investigation that an abandoned county road commission road salt storage site was located near some of these wells (Figure 10).

Well construction records were extracted from the groundwater database and indicated that the glacial drift material was composed of mostly permeable sand and gravel material over a sandstone aquifer (Figure 11). Testing water in the wells compared to the water quality information from the database confirmed that water quality in these high chloride wells was not the norm for this area. The geological cross-section and well-log data also provided the information needed to show how easily wells may become polluted because of a relatively shallow bedrock aquifer with mostly permeable sand and gravel material on top.

Preliminary work such as this became the basis for initializing on-site work to determine the extent of the salt pollution and the steps to be taken to clean the site. In this case, a private groundwater consultant was hired to do a hydrogeological study of the problem, which led to the installation of a groundwater purge system that pumped the salt-polluted water out of both the glacial drift and bedrock aquifer.

4. THE LANDFILL SITE

For many years it seemed convenient and safe to dispose of garbage in holes in the ground, especially when many holes, such as abandoned mines and gravel pits, were so readily available. Regulations were eventually imposed, but some privately owned landfill sites either escaped scrutiny or the regulations came too late. The following example demonstrates how the groundwater database may provide the necessary hydrogeologic information when on-site work may be long in coming.

Many problems connected with groundwater come to the attention of environmental agencies via the expression of concern of the citizens living near the sites generating the problems. The level of concern varies from fear and hysteria to casual curiosity. The bottom line for the citizen is, "this thing is here, what are you going to do about it?" Attempting to ameliorate the situation is irritating, especially when in many cases no provision exists to force a "responsible party" to investigate and correct the problem in a timely fashion. The majority of the time and money is spent on legal matters, which does not serve the best interests of the people affected by the problem or the environment.

Setting these problems aside, can something be accomplished using the

Table 2. Groundwater Chemistry Analysis — Percentiles for SW Michigan Groundwater Project Wells

County: INGHAM
Township(s): ALL
Lithology or Aquifer: ALL

			Percentiles									
Parameter	**N**	**Min**	**1**	**5**	**10**	**25**	**50**	**75**	**90**	**95**	**99**	**Max**
Alkalinity	312	158.0	176.0	196.0	217.0	251.0	286.0	316.0	353.0	384.0	432.0	596.0
Aluminum	312	ND	ND	ND	ND	ND	ND	ND	ND	ND	0.08	0.52
Ammonia	311	ND	ND	0.03	0.04	0.07	0.14	0.21	0.33	0.46	0.76	1.09
Antimony	312	ND	ND	ND	ND	ND	ND	ND	ND	ND	ND	ND
Arsenic	293	ND	ND	ND	ND	ND	ND	0.007	0.013	0.018	0.025	0.042
Barium	312	ND	ND	0.010	0.030	0.070	0.100	0.200	0.400	0.400	0.600	1.400
Boron	312	ND	ND	ND	ND	ND	0.07	0.60	1.80	2.60	3.60	4.70
Cadmium	312	ND	ND	ND	ND	ND	ND	ND	ND	ND	ND	ND
Calcium	312	ND	1.2	8.9	33.0	62.0	76.0	91.0	122.5	156.0	289.0	339.0
Chloride	311	ND	ND	ND	ND	ND	ND	4.0	28.0	52.0	113.0	187.0
Cobalt	312	ND	ND	ND	ND	ND	ND	ND	ND	ND	ND	0.02
Chromium	312	ND	ND	ND	ND	ND	ND	ND	ND	ND	ND	ND
Conductivity	312	319.0	366.0	394.0	426.0	526.0	606.0	703.0	883.0	1124.0	1473.0	1567.0
Copper	312	ND	ND	ND	ND	ND	ND	ND	0.010	0.019	0.120	0.200
Fluoride	303	ND	0.1	0.1	0.1	0.2	0.3	0.6	0.9	1.1	2.2	2.7
Hardness	306	ND	23.0	75.0	160.0	242.0	290.0	350.0	432.0	525.0	649.0	700.0
Iron	312	ND	ND	0.08	0.20	0.40	0.70	1.30	2.60	3.30	7.60	9.20
Iron.2	0	ND	ND	ND	ND	ND	ND	ND	ND	ND	ND	ND
Lead	311	ND	ND	ND	ND	ND	ND	0.002	0.004	0.006	0.009	0.013
Magnesium	312	ND	0.40	2.90	9.50	17.00	24.00	29.00	35.00	40.00	48.00	73.50
Manganese	312	ND	ND	ND	0.01	0.01	0.02	0.04	0.07	0.15	0.60	1.13
Mercury	312	ND	ND	ND	ND	ND	ND	ND	ND	ND	ND	ND
Molybdenum	312	ND	ND	ND	ND	ND	ND	ND	ND	ND	ND	0.01
Nickel	141	ND	ND	ND	ND	ND	ND	ND	ND	ND	ND	ND
Nitrate	312	ND	ND	ND	ND	ND	ND	ND	ND	0.3	0.6	1.3
pH	301	6.10	6.34	6.57	6.66	6.76	6.96	7.37	8.05	8.30	8.41	8.70
Phosphorus	312	ND	ND	ND	ND	ND	ND	ND	ND	0.1	0.2	0.7
Potassium	312	0.49	0.63	0.80	1.00	1.60	3.05	5.40	7.80	8.70	13.00	16.00
Selenium	312	ND	ND	ND	ND	ND	ND	ND	ND	ND	ND	ND
Silica	312	ND	ND	2.90	3.70	4.90	6.60	9.00	17.00	21.00	42.00	48.00
Sodium	312	1.60	2.50	3.20	3.60	4.90	8.40	22.00	76.00	148.00	209.00	313.00
Sodium.2	0	ND	ND	ND	ND	ND	ND	ND	ND	ND	ND	ND
Sulfate	312	ND	ND	1.0	3.0	8.0	19.0	54.5	98.0	148.0	415.0	743.0
Detergent	0	ND	ND	ND	ND	ND	ND	ND	ND	ND	ND	ND
Thallium	312	ND	ND	ND	ND	ND	ND	ND	ND	ND	ND	ND
Zinc	312	ND	ND	ND	ND	0.010	0.040	0.100	0.246	0.400	1.000	2.900

From the Ingham County Health Department, Michigan.

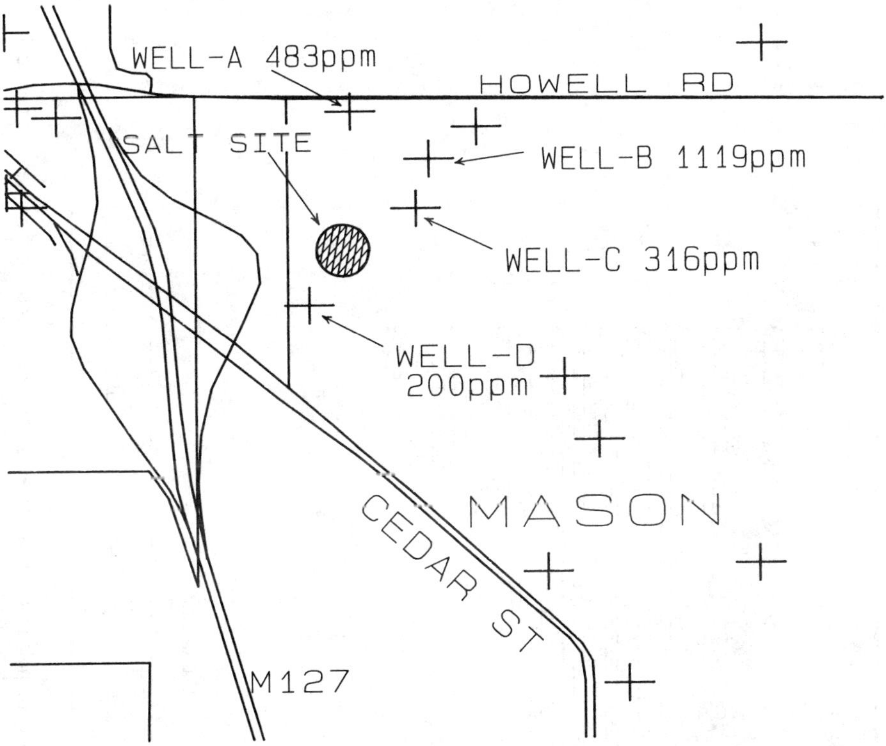

Figure 10. This C-Map graphic indicates the location of a suspected source of salt polllution and several wells in the area. (From the Ingham County Health Department, Michigan.)

groundwater database? The site in question is a private landfill located in Delhi Township, Ingham County, Michigan, which has been closed for several years. A study of the site was started by sampling several privately owned wells near the landfill. A site map is shown in Figure 12. Sample results revealed that water chemistry from several of the wells did not conform to what is usually seen for the area or for the county using STIFF diagrams for comparison (Figure 13). Unusually high levels of sulfate, chloride, sodium, calcium, magnesium, iron, and arsenic were encountered. When the water chemistry results and STIFF diagrams are reviewed in connection with the location of the sampled wells, an obvious pattern is seen as one moves south to north of the landfill. The pattern of affected wells start with Lot-12 near the site, and continues north to Lot-c.

Two geologic cross-sections were made across the landfill (Figures 14 and 15). Figure 14 shows that the landfill may be in permeable soil type material over a layer of shale bedrock. Figure 15 suggests that this shale interface may stop as one moves to the west toward the subdivision. At this point, the permeable soil rests atop a sandstone bedrock. A seepage window may be located at the point the shale

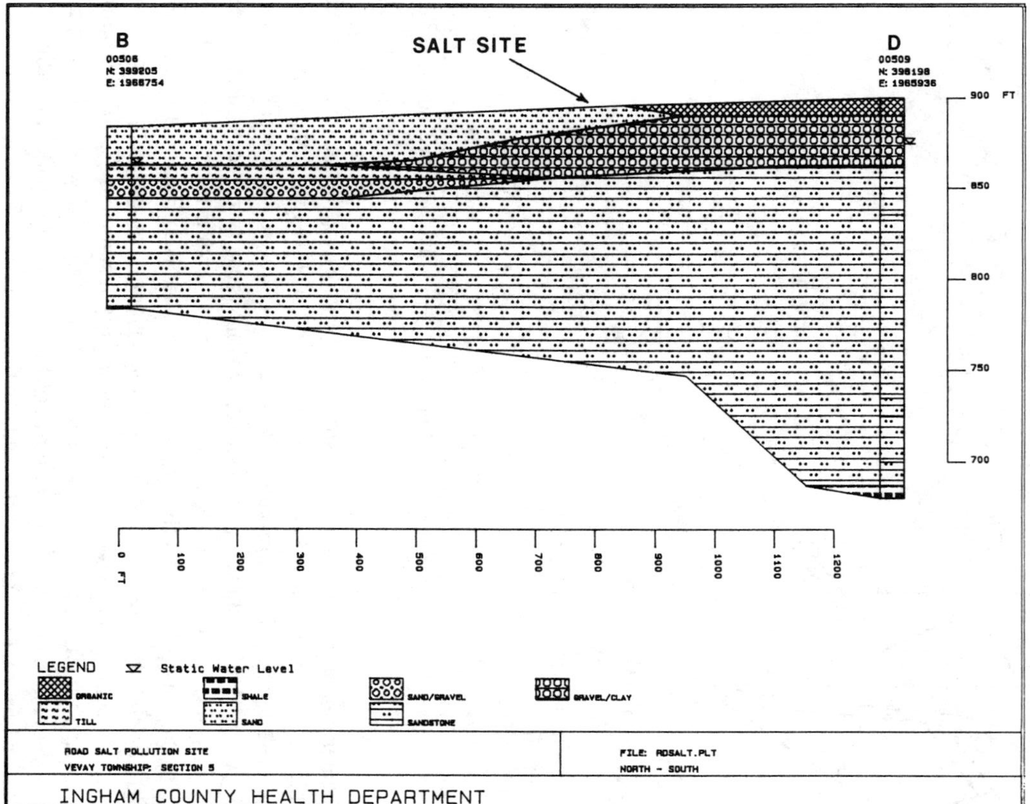

Figure 11. This cross-section, which uses wells B and D from Figure 10, indicates a vulnerable bedrock aquifer due to the extensive permeable layers of sand and gravel on top of the sandstone bedrock. (From the Ingham County Health Department, Michigan.)

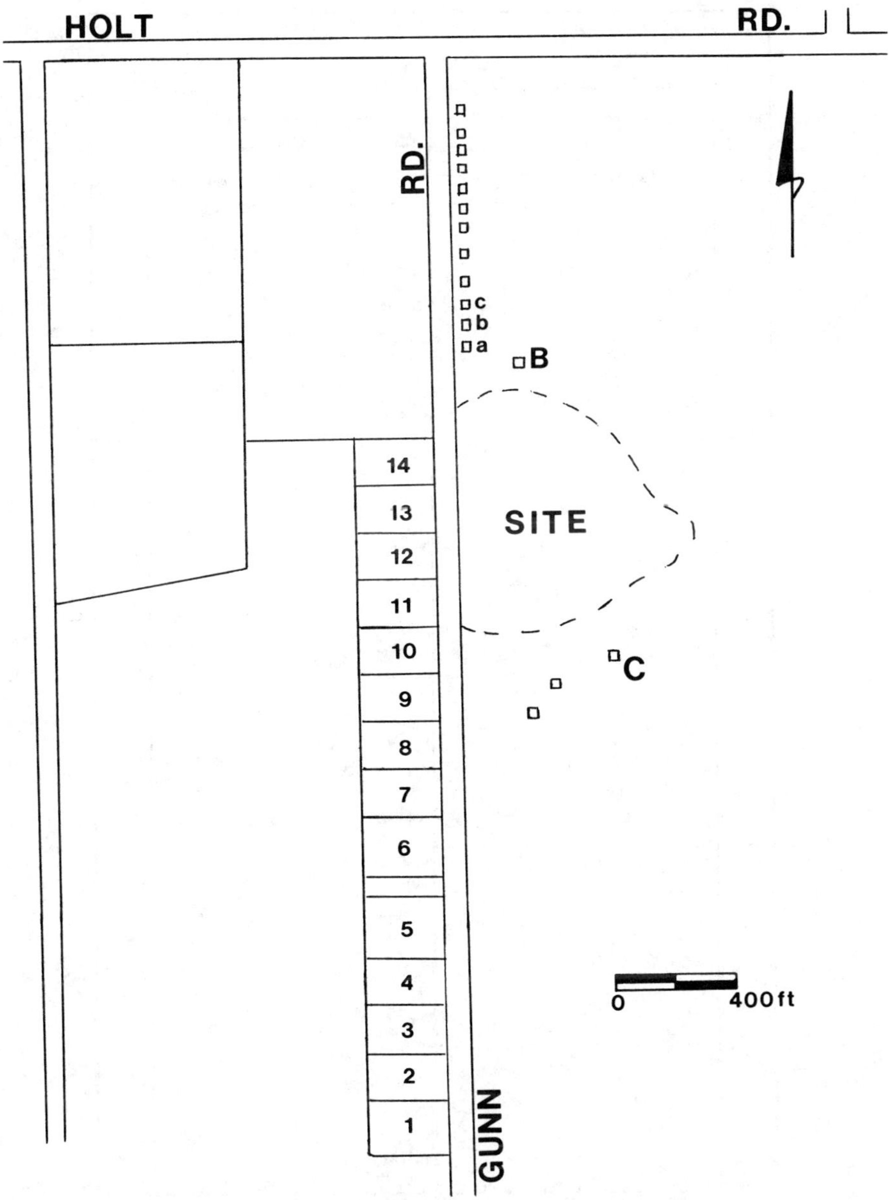

Figure 12. Map showing the locations of several residential home sites to the landfill site. Lots 1 through 14 represent a relatively new subdivision put in several years after the landfill was in operation. The other homes shown just south and north of the site are older homes established while the landfill was in use. (From the Ingham County Health Department, Michigan.)

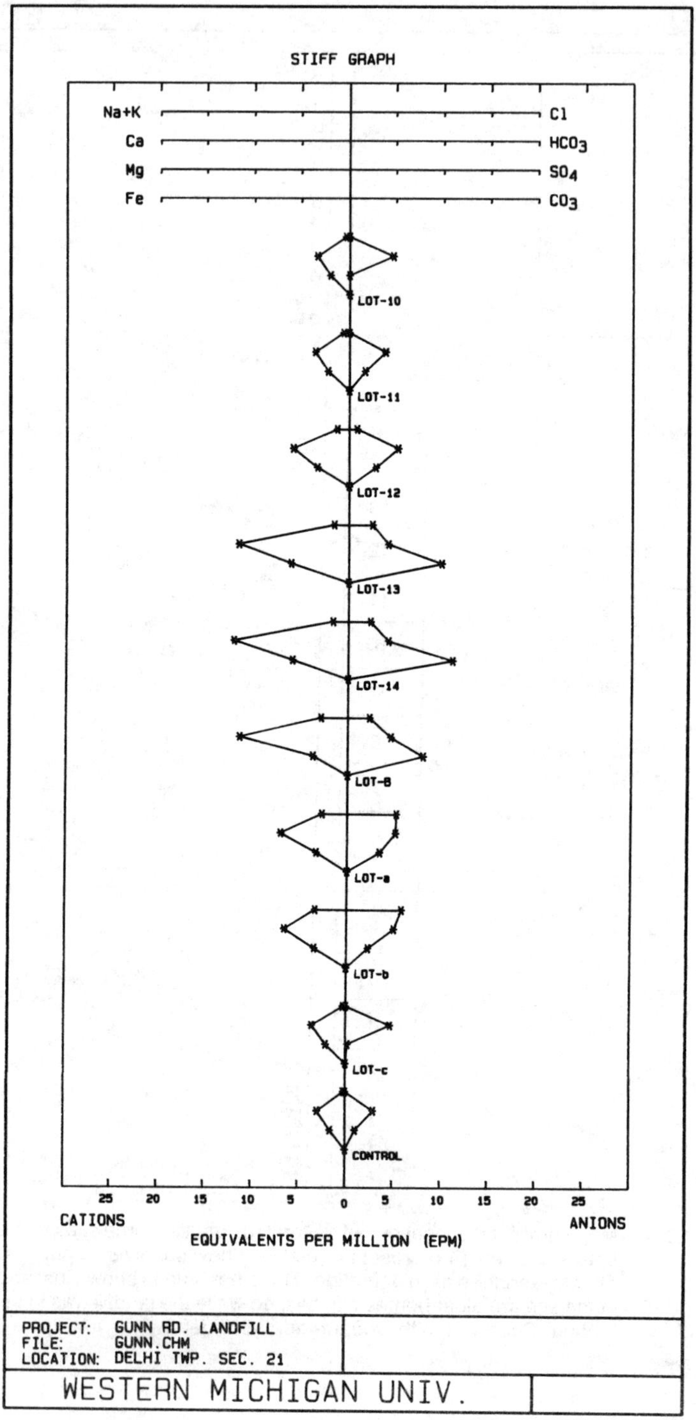
STIFF GRAPH
Na+K
Ca
Mg
Fe
Cl
HCO3
SO4
CO3
LOT-10
LOT-11
LOT-12
LOT-13
LOT-14
LOT-8
LOT-a
LOT-b
LOT-c
CONTROL
25 20 15 10 5 0 5 10 15 20 25
CATIONS
ANIONS
EQUIVALENTS PER MILLION (EPM)
PROJECT: GUNN RD. LANDFILL
FILE: GUNN.CHM
LOCATION: DELHI TWP. SEC. 21
WESTERN MICHIGAN UNIV.

bedrock stops and there is a permeable soil-sandstone bedrock interface. Using this information along with the pattern of affected wells a possible seepage window or recharge zone was mapped (Figure 16).

To further confirm this mechanism, static water level measurements for several sections of the township, which included this site area, were taken from the well construction database. From these data, a potentiometric map of groundwater flow in the bedrock aquifer was made (Figure 17). The direction of groundwater flow in the bedrock was found to be in a NNW direction, consistent with the observed pattern of polluted groundwater. Shallow wells to the south indicate possible southerly leachate movement in a shallow drift aquifer. From this work, a possible plume of polluted groundwater was predicted for the bedrock and shallow drift aquifers (Figure 18).

A third cross-section was made to ascertain what aquifer the subdivision wells were using (Figure 19). The well construction records and cross-section made it obvious that all wells in the subdivision were in the same bedrock aquifer. This helped confirm that affected bedrock wells in this subdivision were not in a different aquifer, which may have been the cause of the different water quality seen.

This type of analysis has limitations. The potentiometric map used static water data from well records that cover several years of drilling. The static water level readings were not measured at the same time and the elevations are based on values from a U.S. Geological Survey topographical map which can have a 10-ft degree of error. These kinds of limitations restrict the database user to a regional rather than a site specific review of groundwater flow. Also, no site information exists for several shallow wells involved, and this makes ascertaining the hydrogeology of the glacial drift layer difficult for the database user. There is still no real substitute for on-site work.

Aside from these limitations, some information regarding the site has recently become available. This includes confirmation that wells in the study area are being polluted, the mechanism by which pollution from the landfill is moving to the affected wells, the direction the plume is taking, and the bedrock aquifer that is affected. This information will help decide the future course of action to be taken to further study the site and provide alternative water to the affected homes, as well as regarding future development near this site. If future facilities fall within a possible zone of influence from this site, considerations will have to be made to protect that water supply. Future action may also include the extension of municipal water to this area or even the restriction of future developments that require on-site wells.

Figure 13. (Opposite page) Water samples taken from home sites near the landfill show a pattern of affected water as one moves from Lot 10 south of the landfill, to the north and past the landfill. Lots 13, 14 and B show the most affected water. These STIFF diagrams for the wells are compared to a "control" well, which reflects unaffected "normal" water for this area. (From the Ingham County Health Department, Michigan.)

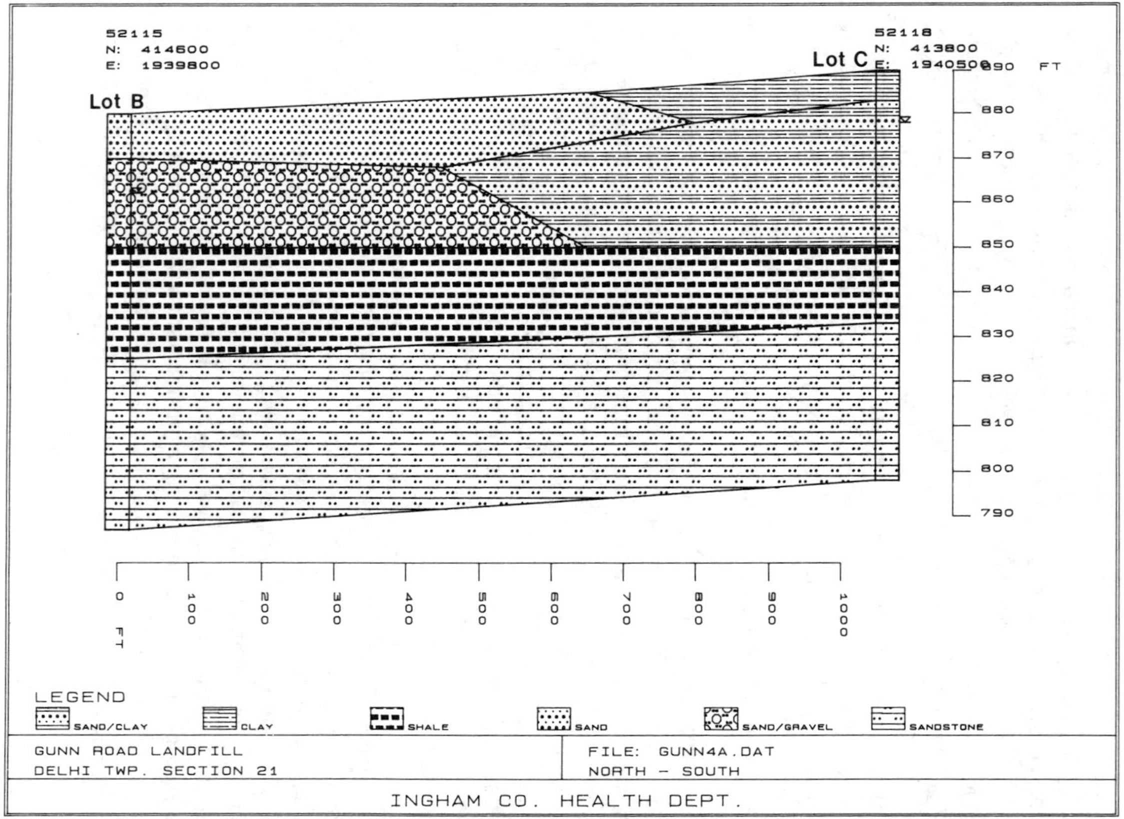

Figure 14. Cross-section using wells from lots B and C. This cross-section indicates some permeable material in the landfill area on top of a relatively impermeable shale bedrock. (From the Ingham County Health Department, Michigan.)

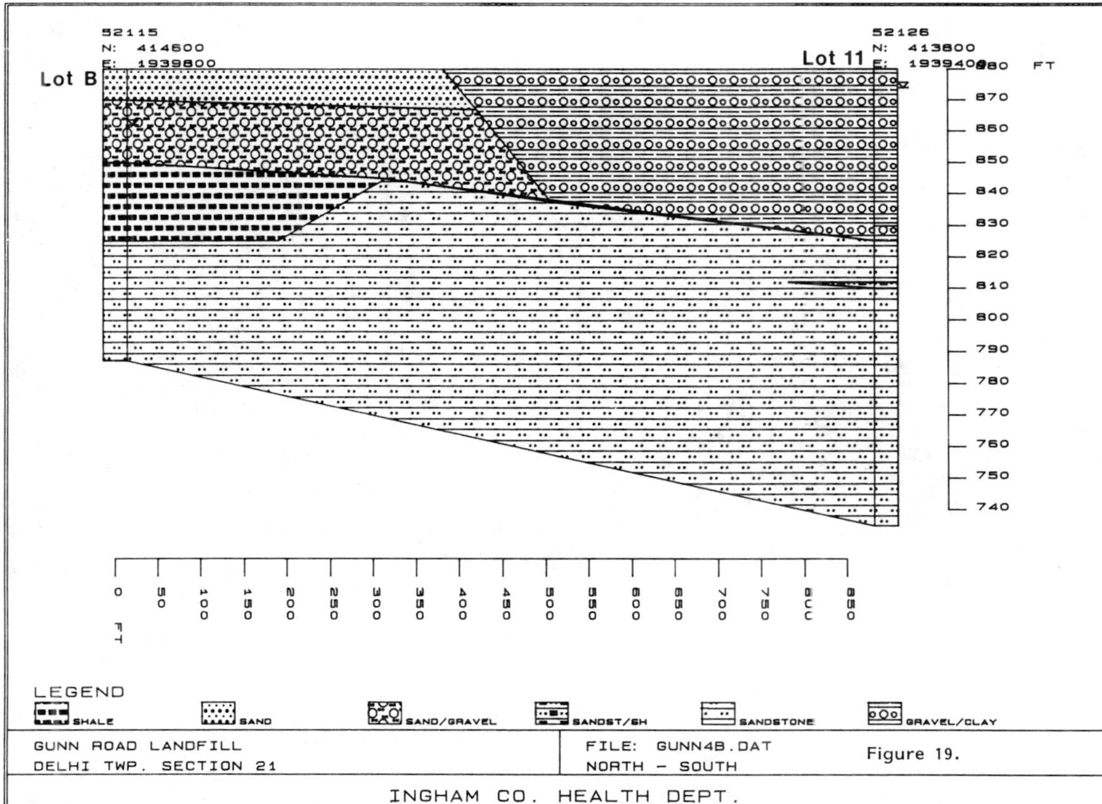

Figure 15. This cross-section using wells at lots B and 11 shows a possible point where permeable sand and gravel material connects with a sandstone bedrock as one moves west from the landfill to the subdivision across from the landfill. Where this connection starts and the shale bedrock tapers off may be a seepage window for landfill leachate to move from the landfill and down into the sandstone bedrock aquifer used by several wells in the area. (From the Ingham County Health Department, Michigan.)

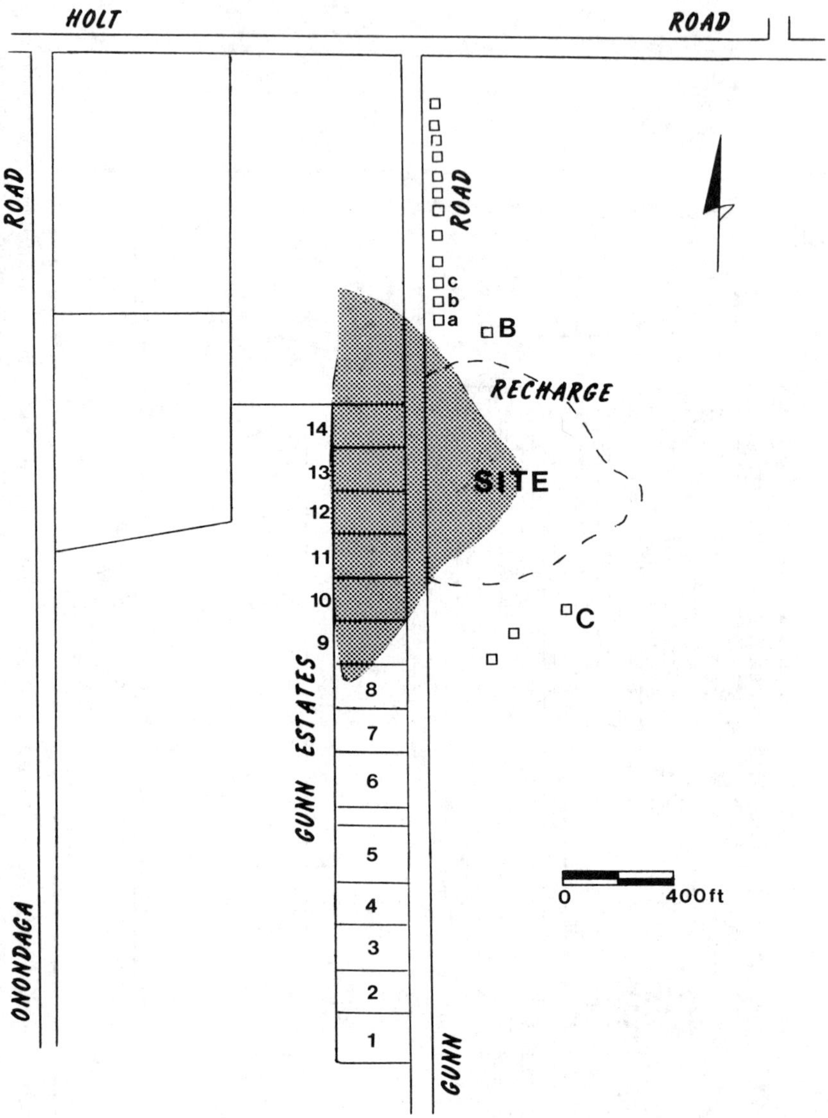

Figure 16. Based on information from the two cross-sections and the pattern of water quality results, a possible seepage or groundwater recharge window was mapped. (From the Ingham County Health Department, Michigan.)

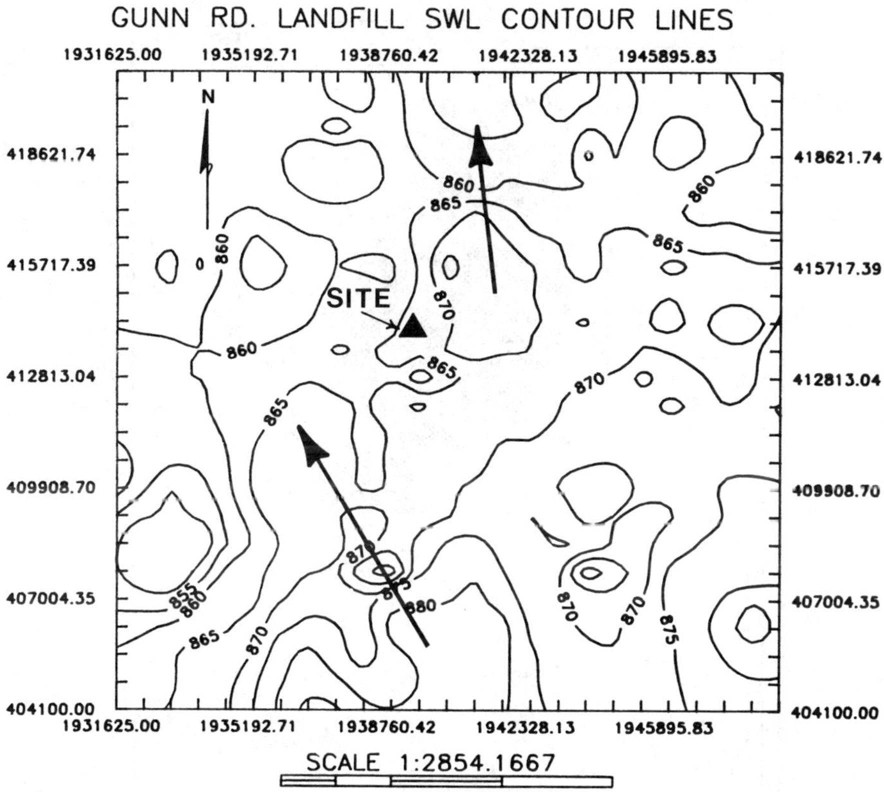

Figure 17. This static water level contour map of the general area around the landfill shows groundwater flow direction in the bedrock aquifer to be in a NNW direction. (From the Ingham County Health Department, Michigan.)

5. WATER CHEMISTRY STUDIES: CORRELATION AND REGRESSION — BORON AND SOFT WATER

Within any water sample are a variety of parameters that can be analyzed. If enough wells are sampled, it may be possible to determine what relationships exist between them. Measurements of relationships are referred to as correlation and regression analyses. A correlation analysis measures a possible relationship between two or more random variables. Correlation does not imply that one variable causes or influences the other and does not look at the relationship between dependent and independent variables. A regression analysis measures the relationship between a dependent and an independent variable or it may measure how one variable is dependent or caused by another independent variable. Correlation may be used initially when one does not have control of either variable. Regression is used primarily when one has experimental control of one variable; however, to the

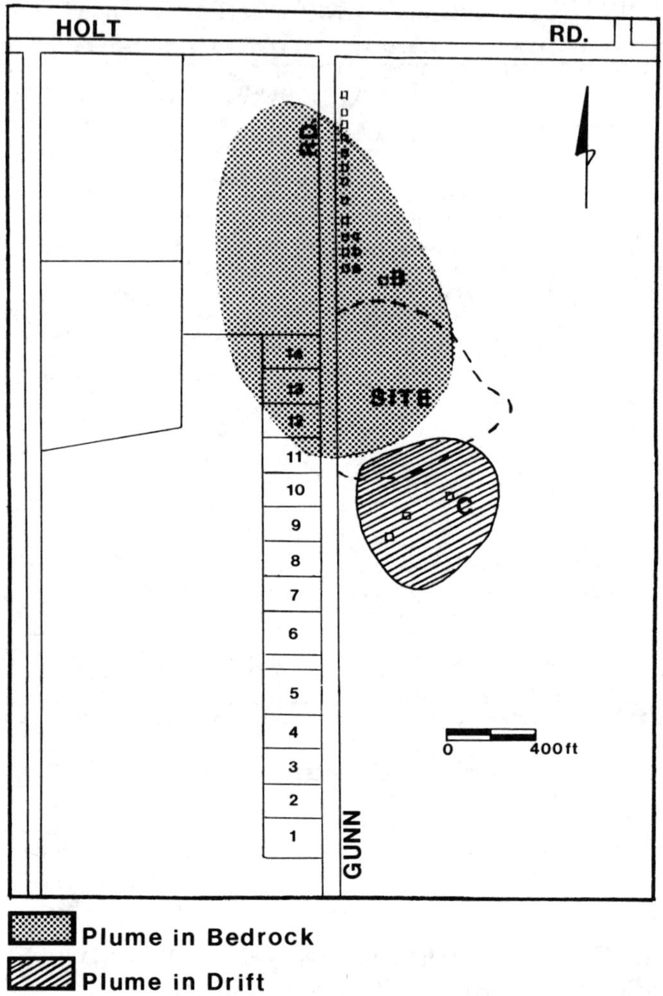

Figure 18. Using the water quality results from wells, the cross-sections, and the static water level contour map, the leachate plume for the bedrock was mapped as moving to the N and NW and included the wells in the shaded area. The smaller plume for the drift was determined again by water test results from shallow wells and speculation that water in the shallow drift aquifer moves toward a low marsh area to the south. (From the Ingham County Health Department, Michigan.)

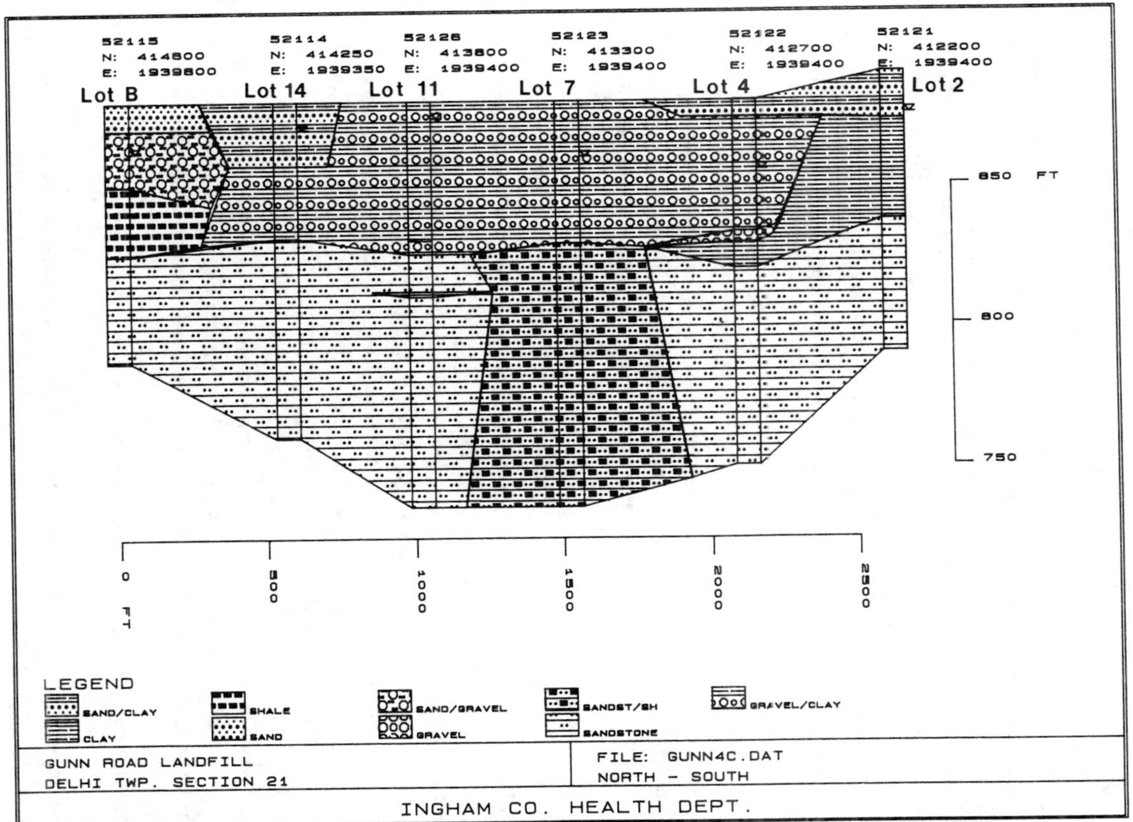

Figure 19. This cross-section includes Lot B and several wells within the subdivision across from the landfill and shows that the wells are using the same bedrock aquifer and that water quality differences are not the result of different aquifer systems. (From the Ingham County Health Department, Michigan.)

objections of some statisticians, regression can be helpful even if one does not have full control of either or both variables if one believes that a causative relationship theoretically exists and one variable influences the other.

Water hardness is made up of several components, including calcium, magnesium, iron, and manganese. If any or all of these components increases, the water hardness also increases. In Figure 20, a simple scatter plot of hardness and magnesium was conducted, and illustrates the relationship of water hardness increasing with higher levels of magnesium. Since this plot suggests a correlation between the two, a correlation analysis was attempted. A Spearman Rank correlation analysis was chosen out of the Statgraphics Software. The results of Table 3 indicate a strong positive correlation between the two. Again, a correlation analysis does not imply that the two variables are caused or affected by the other, only that an apparent relationship exists.

To take this a step further, the database user may conduct a regression analysis for hardness and magnesium. At this point, it may be assumed that a real relationship exists between them and hence see what further may be learned from this analysis. As did the correlation analysis, the regression analysis showed the same significant correlation coefficient (Table 4). The estimated slope indicated that as magnesium increases by 1 unit of measure, hardness should increase approximately 10 units. The regression plot is shown in Figure 21. Statgraphics possesses a convenient feature whereby given any value of the independent variable asked for in the program, a predicted value from the plot for the dependent variable can be reported. Looking at the routine statistics for Ingham County, a percentile chart shows a 50% value of 24.0 ppm for magnesium to a 50% value of 290.0 ppm for hardness (Table 2). When 24.0 ppm is entered for magnesium, the regression plot predicts hardness at 302.84 ppm (Figure 22).

Another example of this analytical tool is a survey of boron levels in Ingham County. The medical community has long been interested in the effects of boron in reducing osteoporosis, a common bone disorder in older women. Some research has shown that a boron supplement of 3 mg/day reduced the urinary excretion of both calcium and magnesium, hence becoming an important factor in reducing the incidence of osteoporosis;[4] however, the U.S. Environmental Protection Agency (EPA) has been reviewing research that may suggest a detrimental effect for boron. Based on available research to date, boron has not been shown to have any carcinogenic or mutagenic potential. However, animal research has shown that boron affects male reproductive organs in both rats and dogs at levels of 23.7 and 29.0 mg/kg/day.[5] A proposed safe level limit for boron in drinking water is 1.0 ppm.

Boron was tested for during the development of a groundwater database in Ingham County. Using the proposed drinking water limit of 1.0 ppm, the water chemistry data can be sorted to obtain those wells with boron levels ≥1.0 ppm.

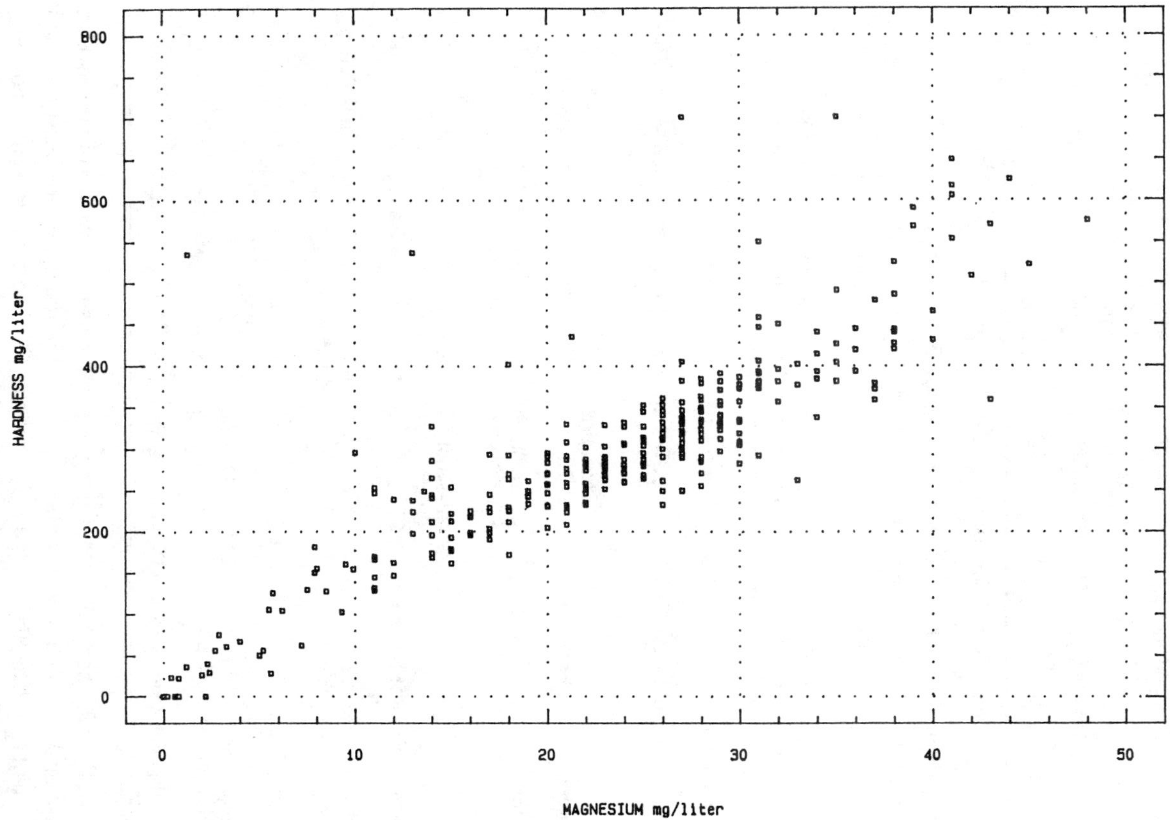

Figure 20. A scatter plot (drawn using the Statgraphics software) of water hardness to water magnesium demonstrating the same positive correlation between the two. With the exception of some data points, much of the data fall on a straight line. (From the Ingham County Health Department, Michigan.)

Table 3. A Spearman Rank Correlation Analysis Performed Out of the Statgraphics Software Showing a Positive Correlation between Water Magnesium and Water Hardness

	Magnesium	Hardness
Magnesium	1.0000	0.8560
	(312)	(312)
	1.0000	0.0000
Hardness	0.8560	1.0000
	(312)	(312)
	0.0000	1.0000

Coefficient (sample size) significance level

From the Ingham County Health Department, Michigan.

Table 4. Calculated Data from the Regression for Water Hardness to Water Magnesium Giving a Positive Correlation Coefficient of 0.805122

Regression Analysis — Linear Model : Y = a + bX

Dependent variable: PROJSTAT.HARDNESS Independent variable: PROJSTAT MAGNESIUM

Parameter	Estimate	Standard Error	T Value	Prob. Level
Intercept	72.5017	10.0615	7.20589	0.00000
Slope	9.41521	0.393933	23.9005	0.00000

Analysis of Variance

Source	Sum of Squares	Df	Mean Square	F-Ratio	Prob. Level
Model	3102327.3	1	3102327.3	571	0.00000
Error	1683583.3	310	5430.9		
Total (Corr.)	4785910.7	311			

Correlation Coefficient = 0.805122 R^2 = 64.82%
Stnd. Error of Est. = 73.6947

From the Ingham County Health Department, Michigan.

These wells can then be plotted geographically using C-Map (Figure 23). A majority of these "high boron wells" are located in the upper northeast quadrant of the county. In Chapter 9, Figure 9, a mapping of "soft water wells" showed a very similar geographic pattern. A map showing both the soft water and the boron wells is shown in Figure 24. Could boron in groundwater be associated with water naturally soft?

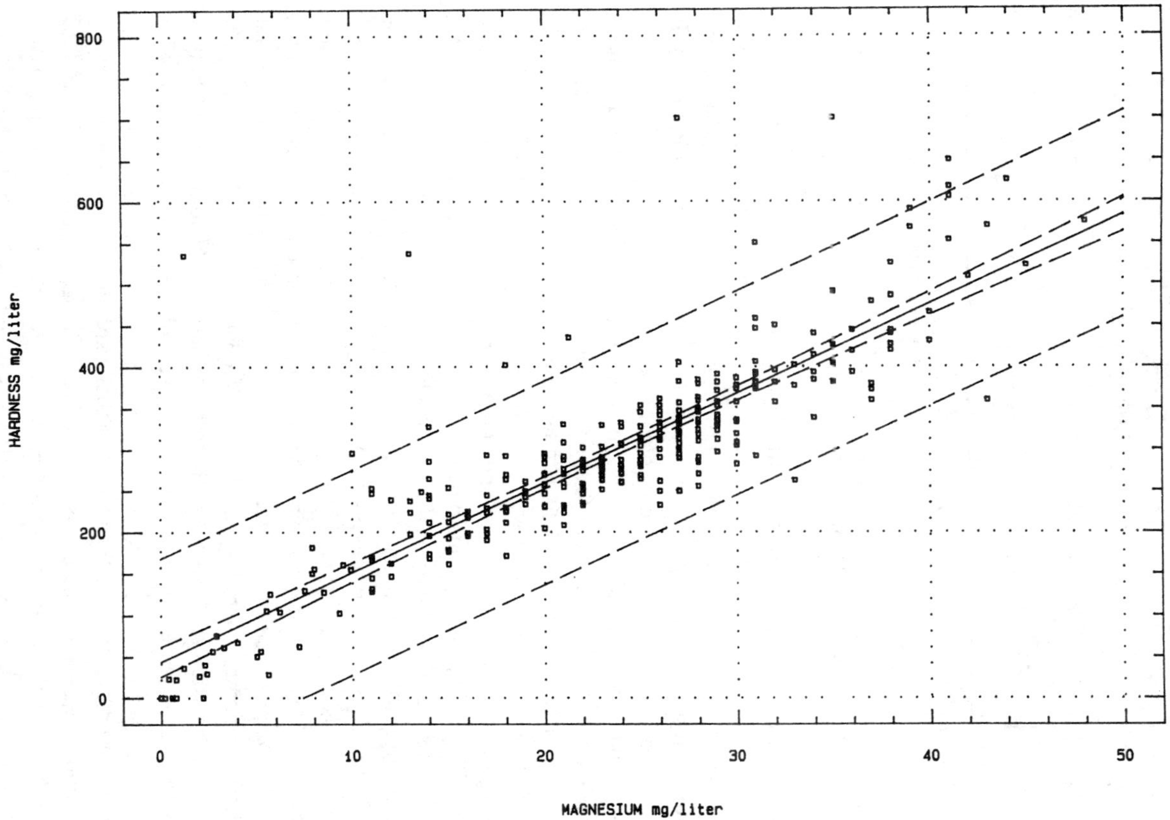

Figure 21. Water quality data taken from five wells in the Ingham County database is used with Hall Groundwater Consultants' software to produce this bar graph using the major cations and anions in the water. (From the Ingham County Health Department, Michigan.) A regression plot shows an approximate 1-to10 relationship between magnesium to hardness. (Estimated slope 9.41).

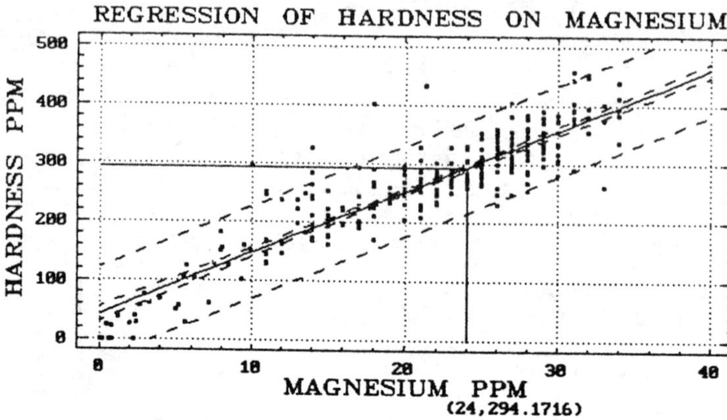

Figure 22. When 24.0 ppm is entered for magnesium, the regression plot predicts hardness at 302.84 ppm.

From the Statgraphics software a simple scatter plot was made for boron vs hardness (Figure 25). A negative correlation is indicated showing that boron increases as hardness decreases. Since sodium is an important parameter associated with soft water another plot of boron and sodium was made (Figure 26). In this case a positive correlation is indicated, as expected.

A Spearman Rank Correlation Analysis was conducted for boron to hardness, calcium, magnesium, and sodium (Tables 5 to 8). As expected, significant but negative correlations were seen between boron and calcium and magnesium and hardness. A positive correlation was found between boron and sodium.

Taking the correlation of boron and sodium, a regression analysis was completed (Table 9). The correlation coefficient of 0.79 is higher than that calculated from the Spearman Rank analysis (0.54). This is due to a difference in calculation, but the results still show a significant positive correlation between sodium and boron. Figure 27 shows the regression plot with a slope estimate of 0.014.

Since boron is not normally analyzed for in wells, a more routine test could help predict a possible boron level that may exceed the proposed drinking water limit of 1.00 ppm being considered by the EPA. Both sodium and hardness are a routine test offered to homeowners by the Michigan Department of Public Health. The results of a regression analysis for both boron to hardness (Table 10) and sodium may signal that boron levels may be too high in a well. Figure 28 is the regression plot.

The sodium regression analysis predicts that a sodium level of 150 will give a boron value of 2.29 and a hardness value of 250 will give a boron value of 0.67 (Figures 29 and 30). The next question is which parameter is better for predicting a boron problem. There are certainly cases where both hardness and sodium are relatively high, especially in situations where salt pollution of a well occurs. A review of the water chemistry database shows 29 of 312 wells with a hardness of >200 ppm and sodium >25 ppm (Table 11). These values for hardness and sodium

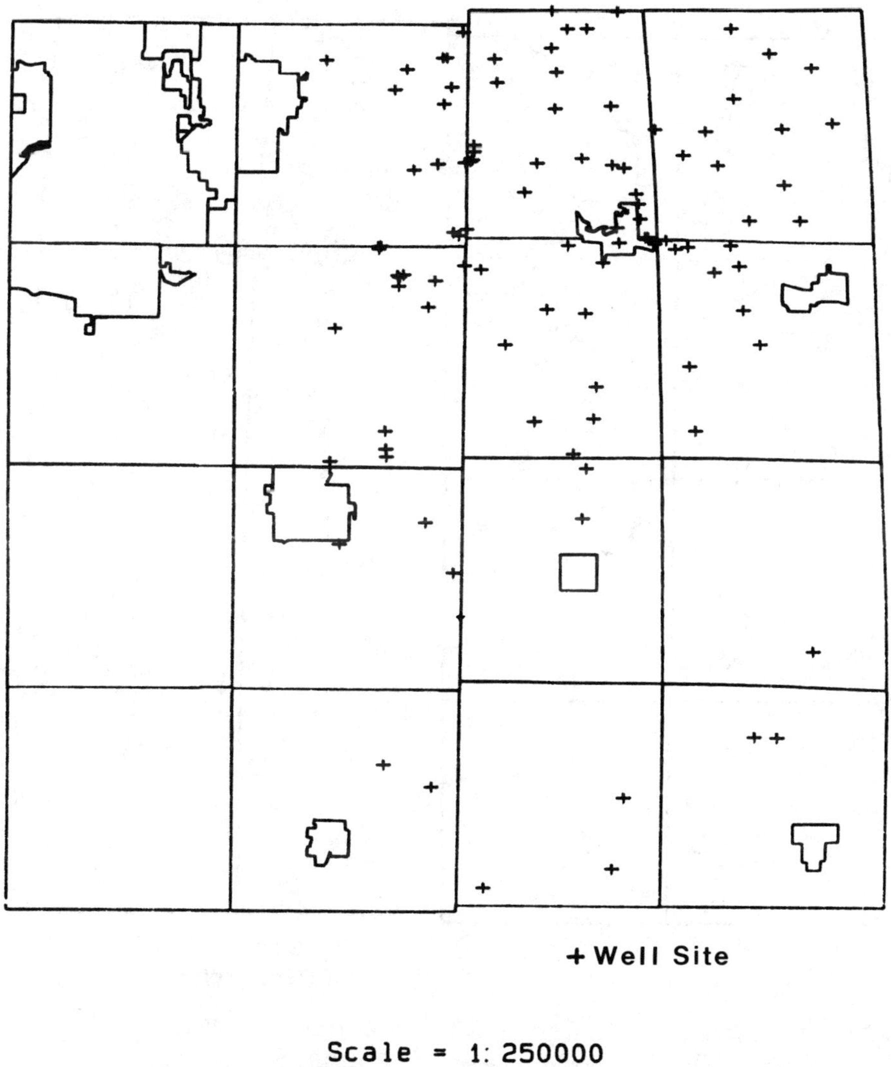

Figure 23. Map showing locations of wells in Ingham County Michigan where boron levels are 1.0 ppm or greater. Most wells are found in the upper NE quadrant. (From the Ingham County Health Department, Michigan.)

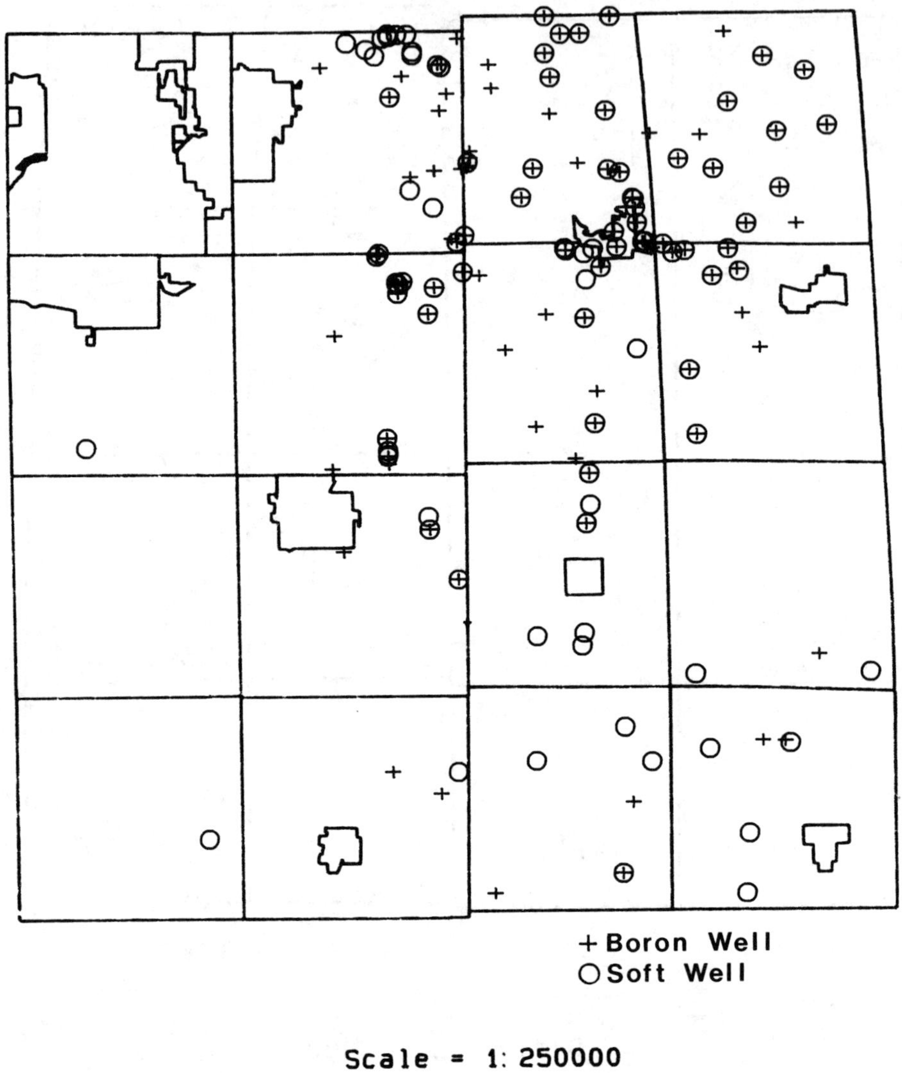

Figure 24. This map shows both "naturally softened wells," (<200.0 ppm hardness as $CaCO_3$) and "high boron wells." Many of the sites are the same, with the locations of both being primarily in the upper NE quadrant. This map suggests that a relationship may exist between boron and "naturally softened water." (From the Ingham County Health Department, Michigan.)

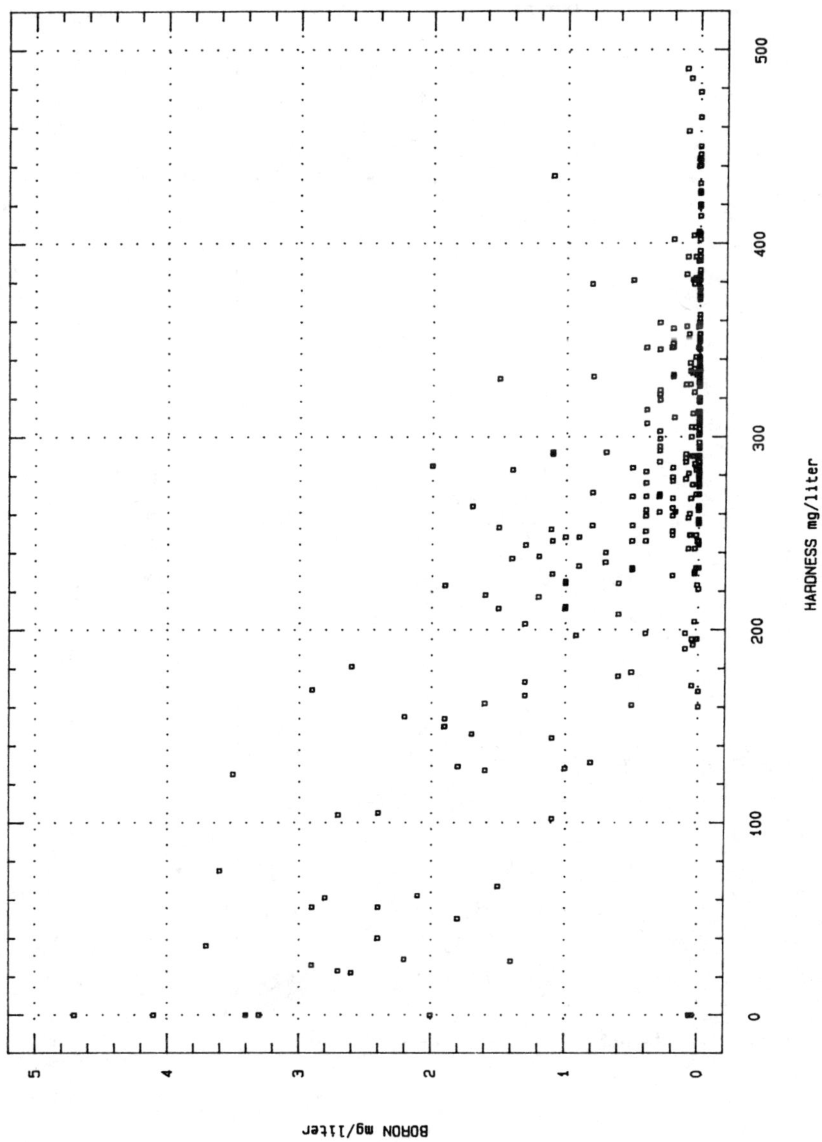

Figure 25. A scatter plot of boron vs hardness from Statgraphics suggesting a negative correlation between boron and water hardness. (From the Ingham County Health Department, Michigan.)

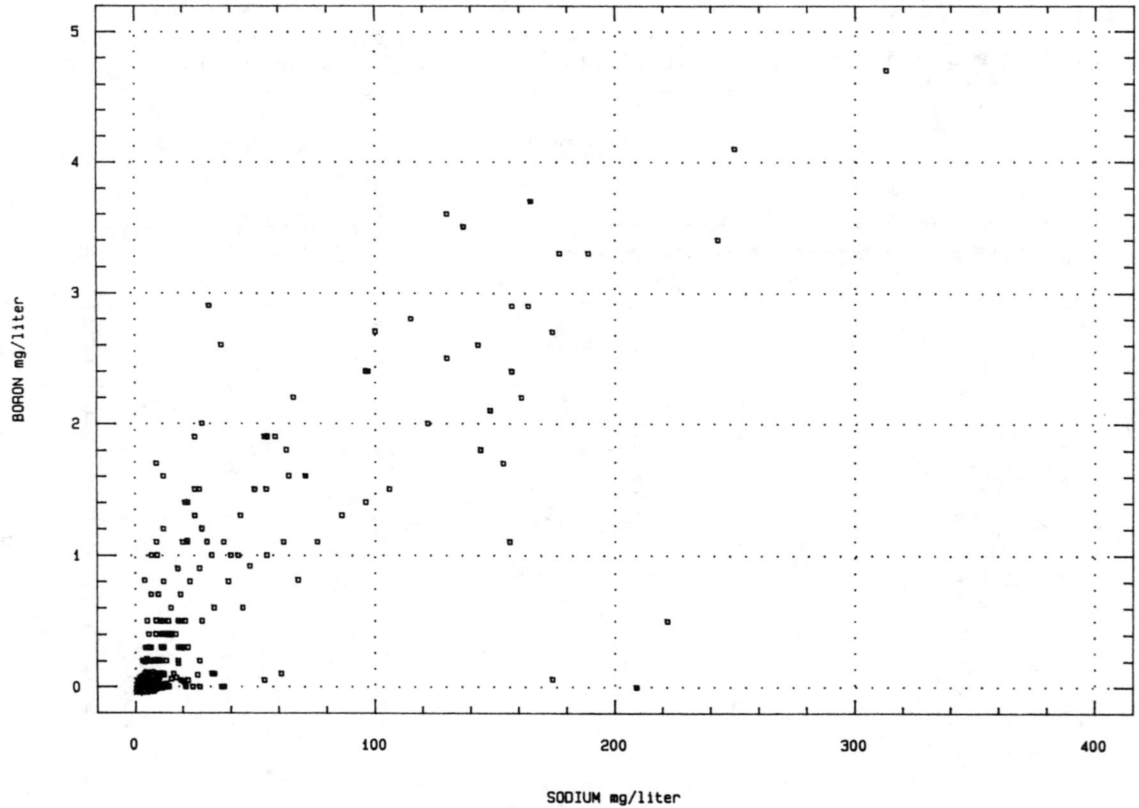

Figure 26. A scatter plot of boron vs sodium from Statgraphics suggesting a positive correlation between boron and sodium. (From the Ingham County Health Department, Michigan.)

Table 5. Spearman Rank Correlation Tests for Boron to Sodium

	Boron	Sodium
Boron	1.0000 (659) 1.0000	0.5446 (659) 0.0000
Sodium	0.5446 (659) 0.0000	1.0000 (659) 1.0000

Coefficient (sample size) significance level

Note: A positive correlation was seen between boron and sodium while negative correlations were seen with the others.

From the Ingham County Health Department, Michigan.

Table 6. Spearman Rank Correlation Tests for Boron to Calcium

	Boron	Calcium
Boron	1.0000 (659) 1.0000	-0.4302 (659) 0.0000
Calcium	-0.4302 (659) 0.0000	1.0000 (659) 1.0000

Coefficient (sample size) significance level

Note: A positive correlation was seen between boron and sodium while negative correlation were seen with the others.

From the Ingham County Health Department, Michigan.

Table 7. Spearman Rank Correlation Tests for Boron to Magnesium

	Boron	Magnesium
Boron	1.0000 (659) 1.0000	-0.5035 (659) 0.0000
Calcium	-0.5035 (659) 0.0000	1.0000 (659) 1.0000

Coefficient (sample size) significance level

Note: A positive correlation was seen between boron and sodium while negative correlation were seen with the others.

From the Ingham County Health Department, Michigan.

Table 8. Spearman Rank Correlation Tests for Boron to Hardness

	Boron	Hardness
Boron	1.0000	-0.4413
	(659)	(659)
	1.0000	0.0000
Hardness	—0.4413	1.0000
	(659)	(659)
	0.0000	1.0000

Coefficient (sample size) significance level

Note: A positive correlation was seen between boron and sodium while negative correlation were seen with the others.

From the Ingham County Health Department, Michigan.

Table 9. Regression Analysis of Boron and Sodium with a Higher Correlation Coefficient Than the One Given from the Spearman Rank (0.792 vs 0.545)

Regression Analysis — Linear Model : Y = a + bX

Dependent variable: PROJSTAT.BORON Independent variable: PROJSTAT.SODIUM

Parameter	Estimate	Standard Error	T Value	Prob. Level
Intercept	0.120341	0.0345299	3.48514	0.00056
Slope	0.0144411	6.3235E-4	22.8371	0.00000

Analysis of Variance

Source	Sum of Squares	Df	Mean Square	F-Ratio	Prob. Level
Model	145.39236	1	145.39236	521.5342	0.00000
Error	86.42124	310	0.27878		
Total (Corr.)	231.81360	311			

Correlation Coefficient = 0.791957 R^2 = 62.72%
Stnd. Error of Est. = 0.527995

From the Ingham County Health Department, Michigan.

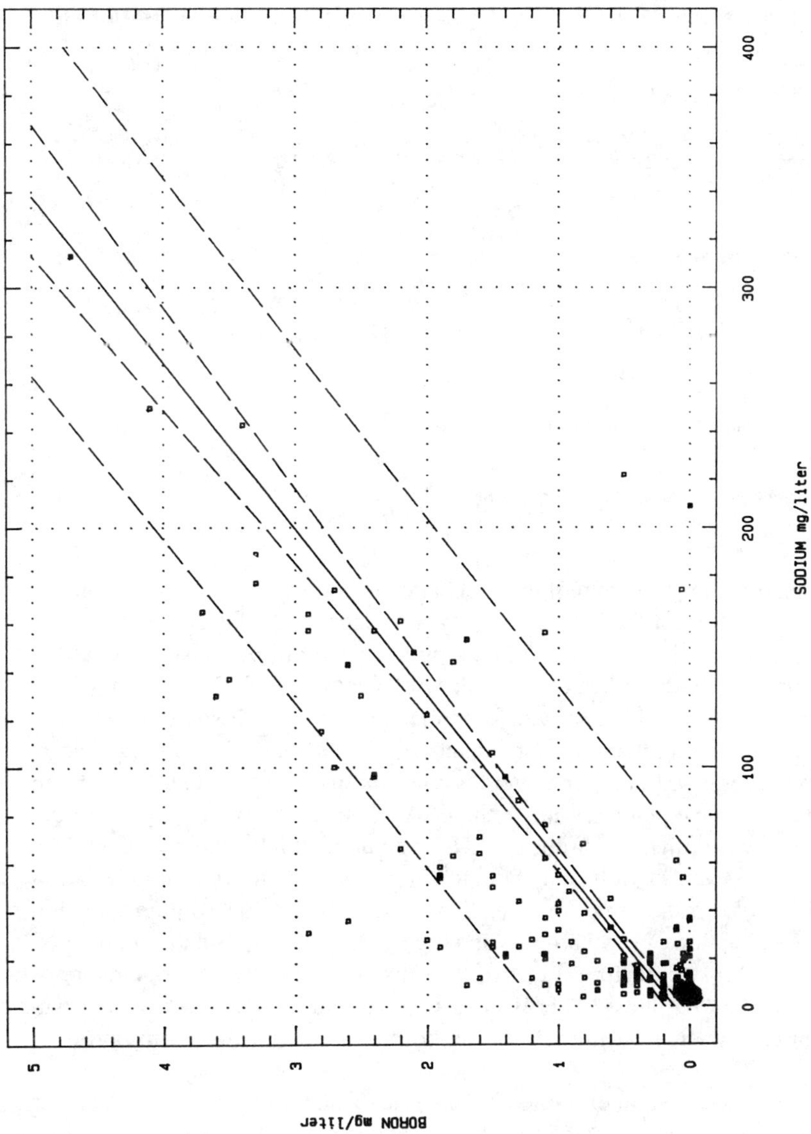

Figure 27. A regression plot of boron on sodium from Statgraphics including the 95% and 90% confidence lines. (From the Ingham County Health Department, Michigan.)

Table 10. A Regression Analysis of Boron and Hardness Indicating the Same Negative Correlation between Them as Shown in the Spearman Rank Analysis

Regression Analysis — Linear Model : Y = a + bX

Dependent variable: PROJSTAT.BORON Independent variable: PROJSTAT.HARDNESS

Parameter	Estimate	Standard Error	T Value	Prob. Level
Intercept	2.22276	0.100489	22.1195	0.00000
Slope	-6.2222E-3	3.44669E-4	-18.0527	0.00000

Analysis of Variance

Source	Sum of Squares	Df	Mean Square	F-Ratio	Prob. Level
Model	118.07118	1	118.07118	325.8985	0.00000
Error	105.78995	292	0.36229		
Total (Corr.)	223.86113	293			

Correlation Coefficient = -0.726244 R^2 = 52.74%
Stnd. Error of Est. = 0.601909

From the Ingham County Health Department, Michigan.

were chosen from a review of past research[5] and the percentile charts and summary statistics[6] completed for the Ingham County water chemistry database (Chapter 9, Tables 1 and 2). Hardness values of 200.0 ppm or less will be considered indicative of naturally soft groundwater, so wells with a hardness of >200.0 ppm will be called high hardness wells, and with sodium >25.0 ppm, high sodium wells.

A review of the Ingham County database shows that 14 wells have high hardness with high sodium levels and high boron levels (Table 12). High boron levels are any value equal to or above the EPA health advisory of 1.0 ppm. Table 13 shows 15 wells with high hardness and high sodium with low boron values. The database also showed 11 wells with both low hardness, low sodium and low boron (Table 14). Table 15 shows no wells that have low hardness, low sodium, and high boron. In the situation of low hardness and high sodium, 37 wells had a high boron level (Table 12). It would appear that sodium may be a slightly better predictor of boron values than hardness. This is consistent with the comparison of the two correlation coefficients obtained from the boron on sodium (0.79), to boron on hardness (-0.73).

As further evidence that exceptions to the rule exist, the database shows that of the water chemistry database of 312, 11 wells showed high hardness, low sodium and high boron (Table 15). Wells in this group (except for one) had a level of potassium that was equal or greater than the average potassium value of 3.8 ppm (Table 16). This situation suggests that potassium may be a factor in boron levels

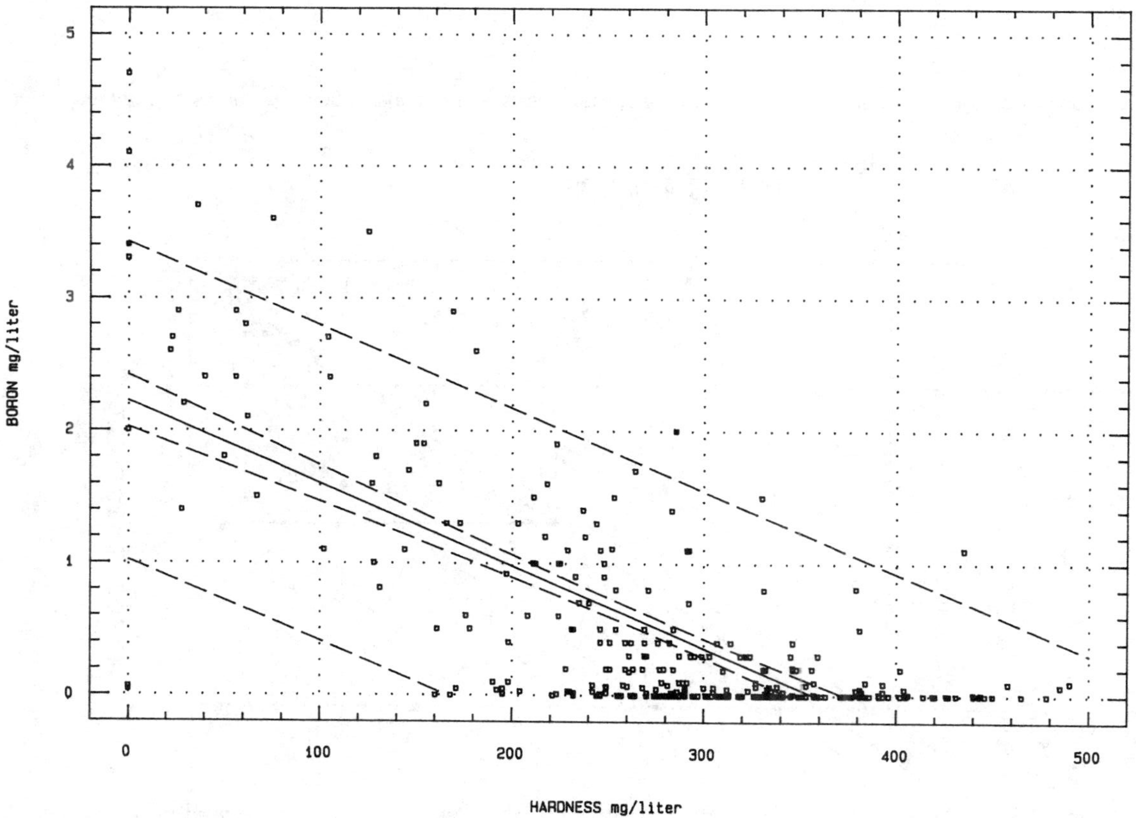

Figure 28. A regression plot of boron on hardness showing the reverse relationship. Can a single parameter such as water hardness help predict a high boron level by the use of a regression plot? Water hardness is often tested for in wells. In the situation where boron is not tested for, other test results may be relied upon for predicting a possible high boron value. (From the Ingham County Health Department, Michigan.)

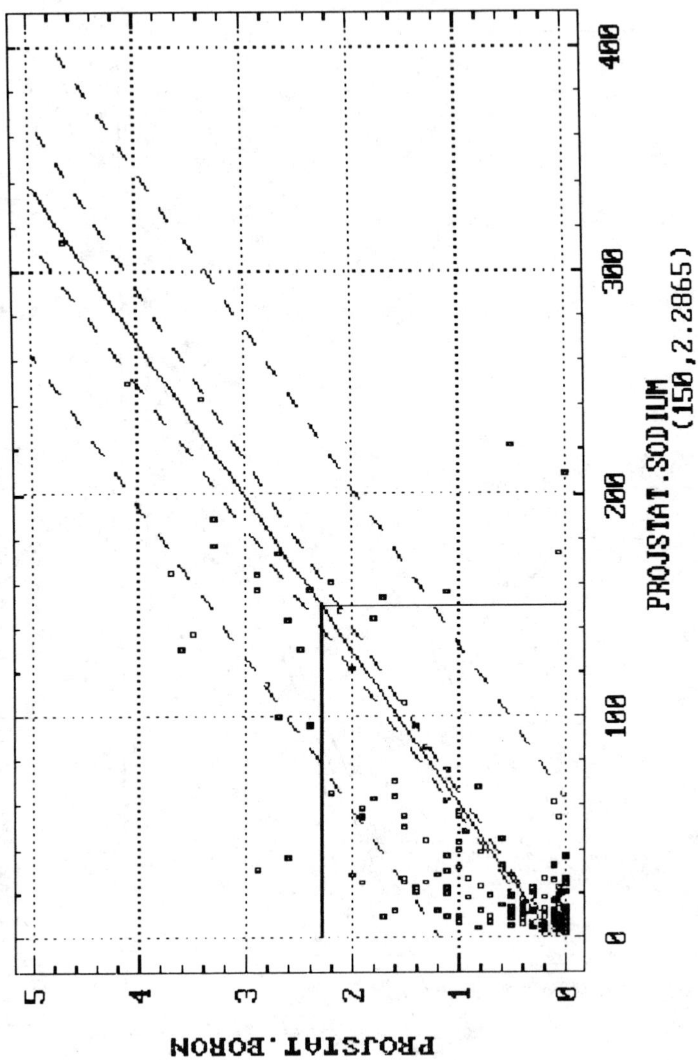

Figure 29. Statgraphics screen showing a regression analysis predicts a boron value of 2.29 in response to a sodium level of 150.

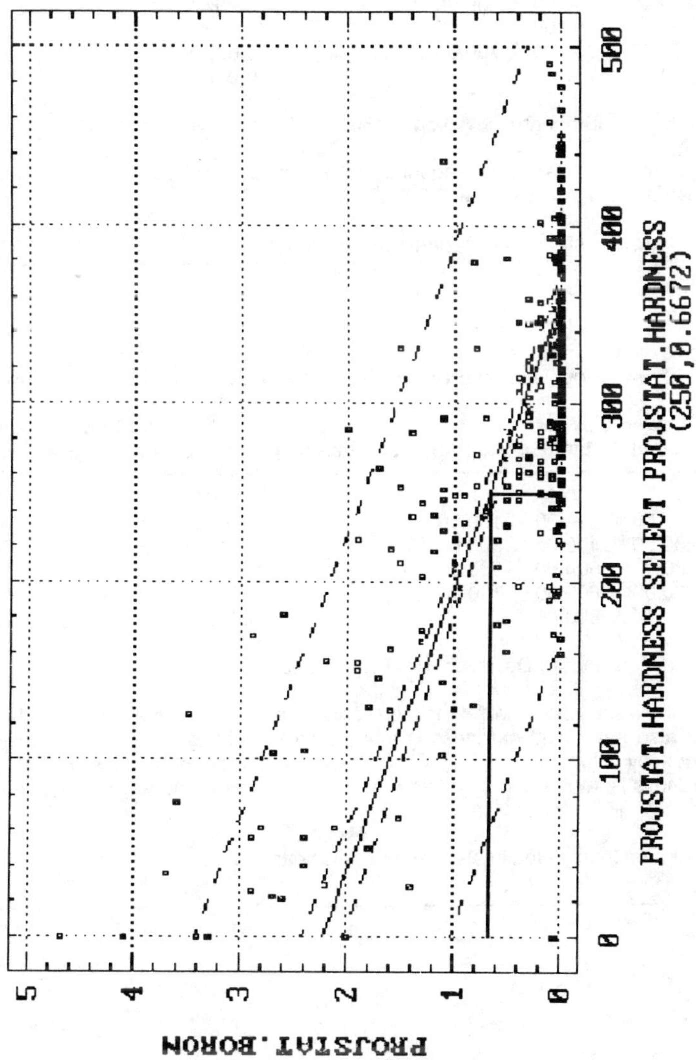

Figure 30. Statgraphics screen showing a regression analysis predicting a boron value of 0.67 in response to a sodium level of 250.

Table 11. Frequency Tabulation for Hardness with Sodium >25 ppm

Class	Lower Limit	Upper Limit	Midpoint	Frequency	Rel. Frequency	Cum. Frequency	Cum. Rel. Frequency
At or below		0.00		7	0.0986	7	0.0986
1	0.00	200.00	100.00	35	0.4930	42	0.5915
2	200.00	400.00	300.00	21	0.2958	63	0.8873
3	400.00	600.00	500.00	6	0.0845	69	0.9718
4	600.00	800.00	700.00	2	0.0282	71	1.0000
Above	800.00			0	0.0000	71	1.0000

Mean = 192.606 Standard Deviation = 160.02 Median = 160

Note: To interpret this table, the frequency column for the condition of sodium >25 rows 3-6 pertain to hardness values≥200 (21 + 6 + 2 + 0).

From the Ingham County Health Department, Michigan.

Table 12. Frequency Tabulation of Hardness with Sodium >25 and Boron >1

Class	Lower Limit	Upper Limit	Midpoint	Frequency	Rel. Frequency	Cum. Frequency	Cum. Rel. Frequency
At or below		0.00		6	0.11760	6	0.118
1	0.00	200.00	100.00	31	0.6078	37	0.725
2	200.00	400.00	300.00	11	0.2157	48	0.941
3	400.00	600.00	500.00	3	0.0588	51	1.000
4	600.00	800.00	700.00	0	0.0000	51	1.000
Above		800.00			0.0000	51	1.000

Mean = 141.137 Standard Deviation = 126.281 Median = 128

Note: This table shows only wells with high sodium and boron, so the number of these wells that also have high hardness is interpreted by adding the last three rows of the frequency column (1 + 3 + 0). Conversely, the number of these wells that have low hardness is interpreted by adding the first two rows of the frequency column (31 + 6).

From the Ingham County Health Department, Michigan.

Table 13. Frequency Tabulation for Hardness with Sodium >25 and Boron <1

Class	Lower Limit	Upper Limit	Midpoint	Frequency	Rel. Frequency	Cum. Frequency	Cum. Rel. Frequency
At or below		0.00		1	0.0500	1	0.0500
1	0.00	200.00	100.00	4	0.2000	5	0.2500
2	200.00	400.00	300.00	10	0.5000	15	0.7500
3	400.00	600.00	500.00	3	0.1500	18	0.9000
4	600.00	800.00	700.00	2	0.1000	20	1.0000
Above	800.00			0	0.0000	20	1.0000

Mean = 323.85 Standard Deviation = 164.718 Median = 329.5

Note: This table shows only wells with high sodium and low boron, so the number of these that also have high hardness is interpreted by adding the last four rows of the frequency column (10 + 3 + 2 + 0).

From the Ingham County Health Department, Michigan.

Table 14. Frequency Tabulation of Hardness with Sodium <25 and Boron <1

Class	Lower Limit	Upper Limit	Midpoint	Frequency	Rel. Frequency	Cum. Frequency	Cum. Rel. Frequency
At or below		0.00		1	0.00442	1	0.00442
1	0.00	200.00	100.00	10	0.04425	11	0.04867
2	200.00	400.00	300.00	183	0.80973	1944	0.85841
3	400.00	600.00	500.00	27	0.11947	221	0.97788
4	600.00	800.00	700.00	5	0.02212	226	1.00000
Above	800.00			0	0.00000	226	1.00000

Mean = 324.336 Standard Deviation = 93.0652 Median = 308

Note: This table shows only wells with low sodium and low boron, so the number of these wells that also have low hardness is interpreted by adding the first two rows of the frequency column (10 + 1).

From theIngham County Health Department, Michigan.

Table 15. Frequency Tabulation of Hardness where Sodium <25 and Boron >1

Class	Lower Limit	Upper Limit	Midpoint	Frequency	Rel. Frequency	Cum. Frequency	Cum. Rel. Frequency
At or below		0.00		0	0.0000	0	0.000
1	0.00	200.00	100.00	0	0.0000	0	0.000
2	200.00	400.00	300.00	10	0.9091	10	0.909
3	400.00	600.00	500.00	1	0.0909	11	1.000
4	600.00	800.00	700.00	0	0.0000	11	1.000
Above	800.00			0	0.0000	11	1.000

Mean = 275.091 Standard Deviation = 90.1193 Median = 248

Note: This table shows only wells with low sodium and high boron, and none of these wells have low hardness. This is interpreted by adding the first two rows of the frequency column (0 + 0). The number of wells that also have high hardness is interpreted by adding the last four rows of the frequency table (10 + 1 + 0 + 0).

From the Ingham County Health Department, Michigan.

Table 16. Listing of Values for each wells with High Hardness, Low Sodium, High Boron, and High Potassium.

Hardness	Sodium	Boron	Potassium
212.000	7.000	1.000	2.900
246.000	9.100	1.100	3.800
252.000	22.000	1.110	4.300
238.000	12.000	1.200	4.600
264.000	9.100	1.700	4.900
248.000	9.300	1.000	5.800
537.000	22.000	1.100	6.000
291.000	20.000	1.100	6.300
237.000	22.000	1.400	6.300
283.000	21.000	1.400	7.600
218.000	12.000	1.600	8.700

Note: From Table 15, which used only wells that had high hardness, low sodium and high boron, and with the exception of one well, had a higher than average potassium value (>3.8). This suggests that potassium is also a factor in boron levels for groundwater.

From the Ingham County Health Department, Michigan.

for groundwater in the rare case where sodium is low and hardness is high. The database user may further explore this by conducting a multiple regression analysis for boron to hardness, sodium, and potassium to see how well a model for boron may fit for these four parameters.

Correlation and regression analyses are useful tools for assisting in research of this type. It is very important that the database user understand the limitations of the statistical tools in relation to his/her understanding of the water chemistry in his/her area. In this example of the boron survey, the statistics cannot provide the evidence that any one of the parameters of hardness, sodium, or potassium has a direct or causative effect on boron. It does suggest that these parameters have a significant relationship to the levels of boron found in groundwater. This kind of information is important in providing a direction for future research.

6. A SUPERFUND SITE

Chapter 1 discussed the necessity of using databases to manage the growing Superfund site list, and emphasized the avoidance of the overlap of effort in the analysis of these sites. This section uses information from the Michigan Statewide Groundwater Database. The creation of these databases on a state level is important because state agencies are usually the group involved with site investigations.

In this description, the actual name of the site is not used. The site is a land disposal site used by a company for some of their wastes. Preliminary work by a consultant determined that organic compounds and metals were found in monitoring wells on-site. A computer-generated site map created by the Michigan Statewide Groundwater Database is shown in Figure 31. This digitized site map provides the locations of monitoring wells, nearby buildings, surface water, and site boundaries.

To emphasize the importance of "databasing" this information, this map can be recreated as an original as many times as needed by accessing the file in the computer and sending it to a plotter. Of more importance is that same digitized map can be used as a base to make further site analysis by adding or subtracting data and information to the map without having to spend money and time on additional drafting and redrawing. To illustrate this, an additional layer of information, such as the static water levels of the monitoring wells, can be retrieved from the database, analyzed by another software package such as the Surfer software demonstrated earlier, and with a GIS software package, simply added to the base map. Figure 32 shows this additional layer of the contours for the static water levels for the glacial drift aquifer. The general flow direction in this aquifer is shown to be to the NNW.

Since the lithology information for the monitoring wells is also included with the database any variety of cross-sections can be created. Figure 33 shows a cross-section originating from monitoring well 16 to monitoring well 1. The cross-section indicates a considerable amount of permeable soil types, sand, and gravel throughout the cross-section. The till material includes any combination of sand, clay, silt, and gravel material.

Figure 31. (See Page 150) A digitized site map provides the locations of monitoring wells, nearby buildings, surface water, and site boundaries. (From the Michigan Statewide Groundwater Database, Michigan Department of Natural Resources.)

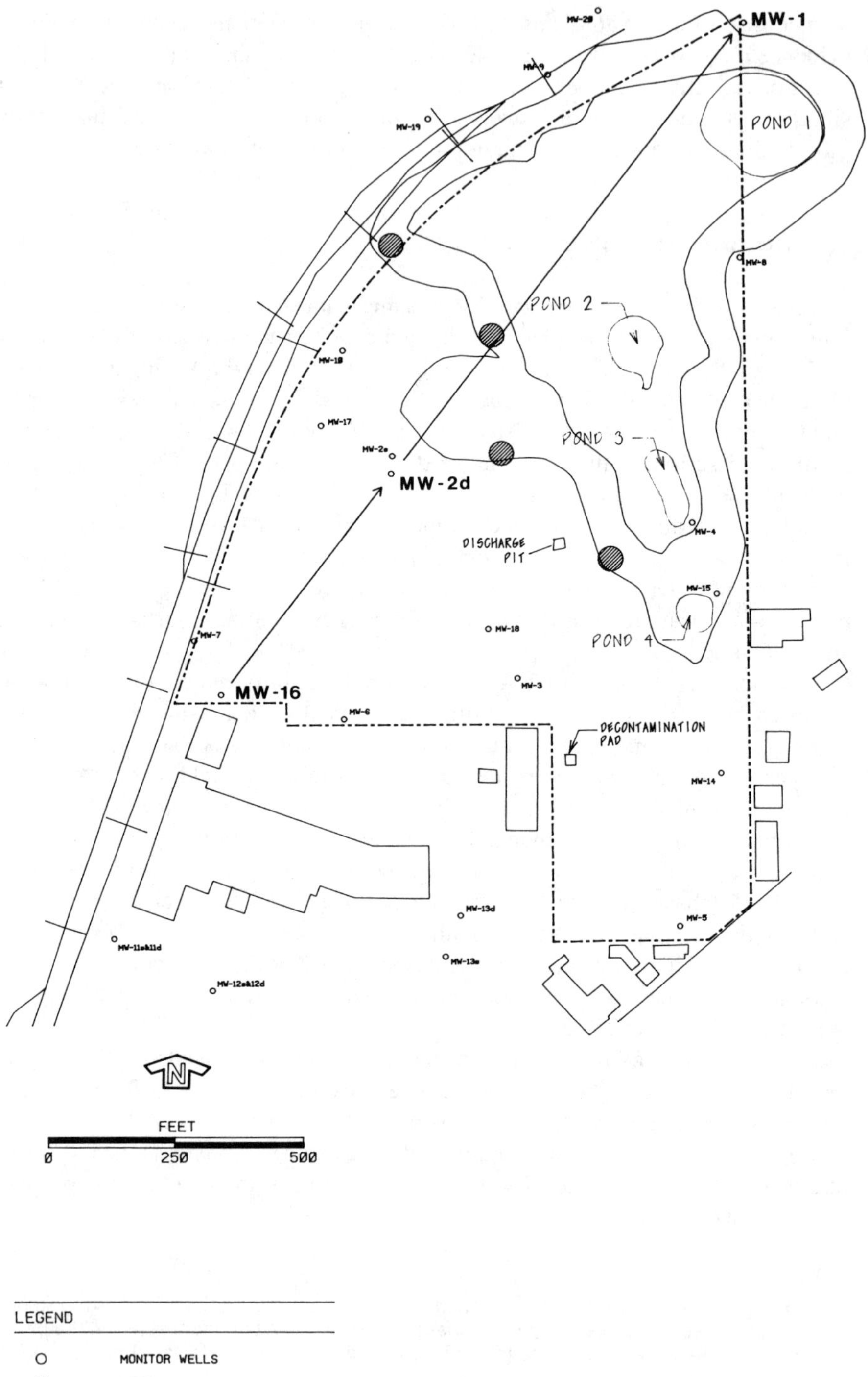
MW-1
POND 1
POND 2
POND 3
POND 4
MW-2d
DISCHARGE PIT
MW-16
DECONTAMINATION PAD
FEET
0
250
500
LEGEND
MONITOR WELLS
SEEPS

Figure 31.

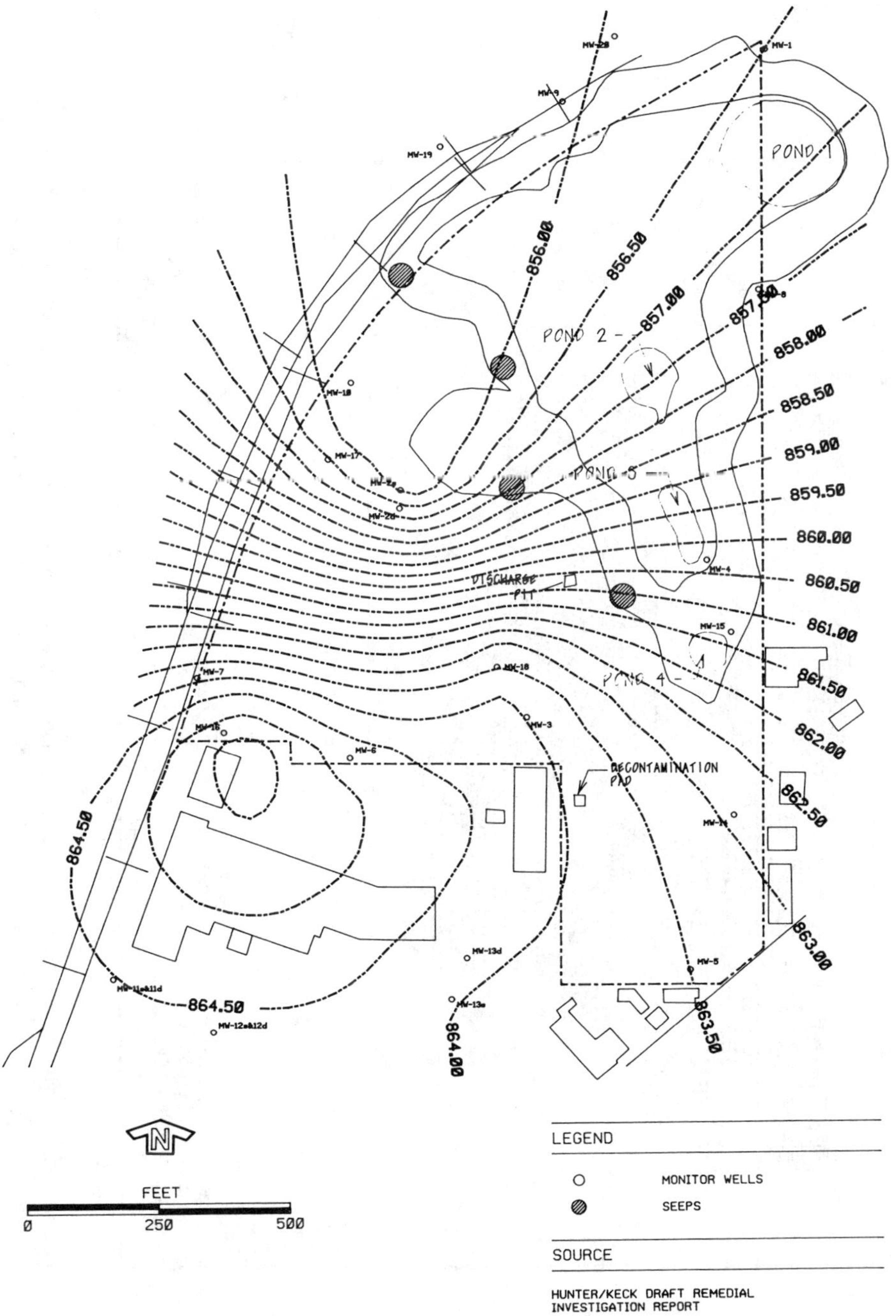

Figure 32. Contour layers for the static water levels for the glacial drift aquifer. The general flow direction in this aquifer is shown to be to the NNW. (From the Michigan Statewide Groundwater Database, Michigan Department of Natural Resources.)

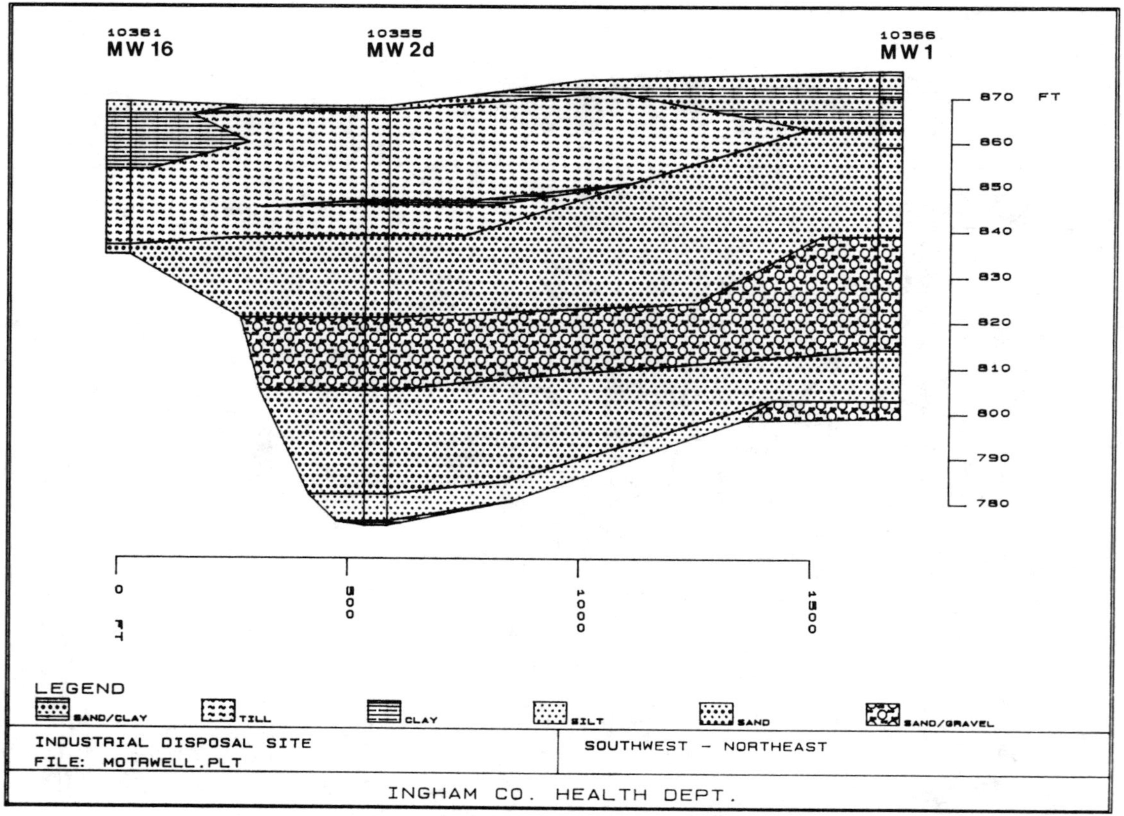

Figure 33. A cross-section originating from monitoring well 16 to monitoring well 1 reveals a considerable amount of permeable soil types sand, and gravel throughout the cross-section. (From the Michigan Statewide Groundwater Database, Michigan Department of Natural Resources.)

If additional consultants, engineers, hydrogeologists, etc., are brought into the project it should not be necessary for the new staff to start from scratch but rather build on the data easily accessible to them through the database. Digitized data and information can be readily communicated through system networks and in a format for GIS application.

7. CONCLUSION

The authors' goal is to show how a computerized management program can provide solutions to the increasing need for groundwater database management and its application to solving pollution problems. An electronic database helps us visualize and communicate information about groundwater resources.

Because this area is still in its infancy, the techniques described in this book for creating a groundwater database are by no means the last word on this topic. Every area of the world is a little different, all with their own specific problems and resources. It is hoped that the process and the examples provided are a good guide to creating a groundwater database and using it effectively.

Many organizations such as the U.S. Environmental Protection Agency and other state agencies are already working on similar projects. Appendix F is a survey from Enslin and Lusch7 showing some federal and state organizations working on computerized groundwater data.

The authors would like to hear from readers involved with computerizing groundwater data or from those who have any questions, comments, or suggestions. Progress in this area will be greatly enhanced as experiences and information are shared.

REFERENCES

1. *C-Map Computer Mapping, Display and Analysis User Manual version 1.03*, Center for Remote Sensing, Michigan State University, East Lansing.
2. Geobase4, Cross-Section Software by Hall Groundwater Consultants. Earthware of California, 30100 Town Center Drive, Suite 0196, Laguna Niguel, CA 92677 (714)495-5727.
3. Surfer, Golden Software, Inc. P.O. Box 281, Golden, CO 80402.
4. Nielson, F. H., C. D. Hunt, L. M. Mullern, and J. R. Hunt. "Effect of Dietary Boron on Mineral, Estrogen, and Testosterone Metabolism in Postmenopausal Women *FASEB J.* 1:394-397 (1987).
5. Weir, R. J., Jr. and R. S. Fisher, "Toxicologic studies on borax and boric acid." *Toxicol. Appl. Pharmacol.*, 23: 351–364 (1972).
6. Vanlier, K. E., W. W. Wood, and J. D. Brunett. Water Supply Development and Management Alternatives for Clinton, Eaton, and Ingham Counties, Geological Survey Water-Supply Paper, Stock No. 2401-02442, prepared in cooperation with the Tri-County Regional Planning Commission and Michigan Department of Natural Resources. U.S. Government Printing Office, Washington, D.C. (1969).

7. Stoline, M. R., A. E. Kehew, R. N. Passero, P. S. Miller, and G. W. Rowe. "Groundwater Chemistry Statistical Summaries for Ingham County, Michigan", in *Report on the Aquifers of Ingham County: A Survey of Groundwater Quality in the Saginaw Formation for Ingham County,* Rowe, G., Ed. 1989.
8. Enslin W. R., and D. P. Lusch. National Inventory of Statewide Groundwater Data Bases, Part of Systems Analysis and Design of a Statewide Groundwater Data Base, Michigan State Univesity, prepared for Land and Water Management Division, Michigan Department of Natural Resources (1987).

Key Words

Aeration
Aerial
Agencies
Agencies
Alkalinity
Anion
Aquifer
Aquipro
Arsenic
Aubbeenaubbee-Capac
Average
Background water quality
Bacteriological
Bar graphs
Baseline
Bedrock
Bentonite
Boron
C-Map
Calcium
Cations
Chemical tanks
Chemistry
Chloride
Clay
Coliform bacteria
Colwood-Brockston
Computers
Conductivity
Contaminated
Contour maps
Coordinates
Correlation
County
Cross-section
Database unit
Database software
Database file structure
Database error checking
Database attributes
dBase III
Decision making
Degree of error
Digitizing
Discharge
Drift
Drinking water
EPA
Fence diagram
Filters
Flow patterns
Foxbase
Frequency table
Fuel tanks

Scatter plot
Sections
Seepage window
Septic tank
Sewage, wastewater
Sewage, septic
Sewage
Shale
Site investigations
Sodium
Soft
Software
Soil
Spearman Rank Correlation
Standards
State plane coordinates
Statgraphics
Static water level
Statistics
STIFF Diagram
Streams
Subsurface
Sulfate
Surface
Surfer
Surveys
Tanks
Tax ID
Tax map
Tax rolls
TDS
Temperature
Tier range
Topographic
Total dissolved solids (TDS)
Townships
Townships, decision making
Transition
Underground tanks
Underground storage
Unprotected
User
U.S. Geological Survey (USGS)
UTM
Vulnerable
Wastewater
Wastewater leachage
Wastewater lagoon
Water quality
Water softener
Water filters
Well ID number
Well injection
Well monitoring
Well owner
Well driller
Well casing
Well construction
Well
Well grout
Well flushing
Well logs
Well location
Well permits
Well accuracy
Wellkey
Zone of influence

Appendix A

State Superfund Offices (by EPA Regions)

Region 1

Connecticut

Department of Environmental Protection
Hazardous Materials Management Unit
165 Capitol Avenue
Hartford, CT 06106
(203) 566-4633

Maine

Department of Environmental Protection
Bureau of Oil and Hazardous Materials Control
Division of Licensing and Enforcement
State House - Station 17
Augusta, ME 04333
(207) 289-2651

Massachusetts

Department of Environmental Quality Engineering
Division of Solid and Hazardous Waste
One Winter Street
5th Floor
Boston, MA 02108
(617) 292-5648

New Hampshire

Department of Environmental Services
Division of Waste Management
6 Hazen Drive
Concord, NH 03301-6509
(603) 271-2908

Rhode Island

Department of Environmental Management
Solid Waste Management Program
Cannon Building, Room 204
75 Davis Street
Providence, RI 02908
(401) 277-2797

Vermont

Agency of Natural Resources
Department of Environmental Conservation
Waste Management Division
West Building
103 South Main Street
Waterbury, VT 05676
(802) 244-8702

Region 2

New Jersey

Department of Environmental Protection
Division of Hazardous Site Mitigation
401 East State Street
CN 028
Trenton, NJ 08625
(609) 984-2902

New York

Department of Environmental Conservation
Division of Solid and Hazardous Waste
Bureau of Resource Recovery
50 Wolf Road
Room 208
Albany, NY 12233
(518) 457-7337

Puerto Rico

Puerto Rico Environmental Emergency Commission
Environmental Quality Board
P.O. Box 11488
Santurce, PR 00910-1488
(809) 722-0077

Region 3

Delaware

Department of Natural Resources and Environmental Control
Division of Air and Waste Management
P.O. Box 1401
89 Kings Highway
Dover, DE 19903
(302) 736-3672

District of Columbia

Consumer and Regulatory Affairs
Pesticides and Hazardous Waste Management
5010 Overlook Avenue, S.W.
Room 114
Washington, D.C. 20032
(202) 783-3194

Maryland

Department of the Environment
Waste Management Administration
Office of Environmental Programs
201 West Preston Street
Baltimore, MD 21201
(301) 225-6953

Pennsylvania

Department of Environmental Resources
Bureau of Waste Management
P.O. Box 2063
Fulton Building
8th Floor
Harrisburg, PA 17120
(717) 783-7816

Virginia

Department of Waste Management
Monroe Building
11th Floor
101 North 14th Street
Richmond, VA 23219
(804) 225-2667

West Virginia

Department of Natural Resources
Division of Waste Management
1260 Greenbrier Street
Charleston, WV 25311
(304) 348-5953

Region 4

Alabama

Department of Environmental Management
Land Division
1751 Federal Drive
Montgomery, AL 36130
(205) 271-7730

Florida

Department of Environmental Regulation
Bureau of Waste Management
2600 Blair Stone Road
Tallahassee, FL 32399-2400
(904) 488-0190

Georgia

Department of Natural Resources
Land Protection Branch
Industrial and Hazardous Waste Management Program
205 Butler Street, S.E.
Room 1154
Atlanta, GA 30334
(404) 656-7404

Kentucky

Department for Environmental Protection
Division of Waste Management
18 Reilly Road
Frankfort, KY 40601
(502) 564-6716

Mississippi

Department of Natural Resources
Bureau of Pollution Control
Division of Solid Waste Management
P.O. Box 10385
Jackson, MS 39209
(601) 961-5062

North Carolina

Department of Human Resources
Solid and Hazardous Waste Management Branch
P.O. Box 2091
Raleigh, NC 27602
(919) 733-2178

South Carolina

Department of Health and Environmental Control
Bureau of Solid and Hazardous Waste Management
2600 Bull Street
Columbia, SC 29201
(803) 734-5200

Tennessee

Division of Solid Waste Management
Customs House
4th Floor
701 Broadway
Nashville, TN 37219-5403
(615) 741-6287

Region 5

Illinois

Environmental Protection Agency
Division of Land Pollution Control
P.O. Box 19276
2200 Churchill Road
Springfield, IL 62794-9276
(217) 782-6760

Indiana

Department of Environmental Management
Office of Solid and Hazardous Waste
105 South Meridian Street
Indianapolis, IN 46225
(317) 232-8941

Michigan

Department of Natural Resources
Emergency Response Division
P.O. Box 30028
Lansing, MI 48909
(517) 373-4823

Minnesota

Pollution Control Agency
Division of Solid and Hazardous Waste
520 Lafayette Road, North
St. Paul, MN 55155
(612) 296-7290

Ohio

Environmental Protection Agency
Division of Solid and Hazardous Waste Management
P.O. Box 1049
1800 Watermark Drive
Columbus, OH 43266-1049
(614) 481-7200

Wisconsin

Department of Natural Resources
Bureau of Solid Waste Management
101 South Webster Street
Madison, WI 53707
(608) 267-7562

Region 6

Arkansas

Department of Pollution Control and Ecology
Solid Waste Division
P.O. Box 9583
8001 National Drive
Little Rock, AR 72209
(501) 562-7444

Louisiana

Department of Environmental Quality
Office of Solid and Hazardous Waste
P.O. Box 44307
Baton Rouge, LA 70804
(504) 342-8925

New Mexico

Health and Environmental Department
Solid and Hazardous Waste Management Program
P.O. Box 968
Santa Fe, NM 87504-0968
(505) 827-2775

Oklahoma

Department of Health
Waste Management Service
Solid Waste Division
P.O. Box 53551
1000 N.E. 10th Street
Oklahoma City, OK 73152
(405) 271-5338

Texas

Water Commission
Division of Hazardous and Solid Waste
Superfund Section
P.O. Box 13087
Capital Station
Austin, TX 78711
(512) 463-7785

Region 7

Iowa

Department of Water, Air and Waste Management
Henry A. Wallace Building
900 East Grand
Des Moines, IA 50319-0034
(515) 281-4968

Kansas

Department of Health and Environment
Bureau of Waste Management
Forbes Field
Topeka, KS 66620
(913) 296-1675

Missouri

Department of Natural Resources
Waste Management Program
Jefferson Building
P.O. Box 176
205 Jefferson Street
Jefferson City, MO 65102
(314) 751-3176

Nebraska

Department of Environmental Control
Land Quality Division
P.O. Box 94877
Lincoln, NE 68509-4877
(402) 471-4217

Region 8

Colorado

Department of Health
Hazardous Materials and Waste Management Division
4210 East 11th Avenue
Denver, CO 80220
(303) 331-4830

Montana

Department of Health and Environmental Sciences
Solid and Hazardous Waste Bureau
Cogswell Building
Room B-201
Helena, MT 59620
(406) 444-2821

North Dakota

Department of Health
Division of Hazardous Waste Management and Special
Studies
P.O. Box 5520
1200 Missouri Avenue
Room 302
Bismarck, ND 58502-5520
(701) 224-2366

South Dakota

Department of Water and Natural Resources
Office of Air Quality and Solid Waste
Joe Foss Building
Room 217
523 East Capitol
Pierre, SD 57501
(605) 773-3153

Utah

Bureau of Solid and Hazardous Waste
Division of Environmental Health
P.O. Box 16690
288 North 1460 West
Salt Lake City, UT 84116-0690
(801) 538-6170

Wyoming

Department of Environmental Quality
Solid Waste Management Program
Herscler Building
122 West 25th Street
Cheyenne, WY 82002
(307) 777-7752

Region 9

American Samoa

Environmental Quality Commission
American Samoa Government
Pago Pago, AS 96799
(684) 633-2304

Arizona

Department of Health Services
Division of Environmental Health Services
1740 West Adams Street
Phoenix, AZ 85007
(602) 257-2306

California

Department of Health Services
Toxic Substances Control Division
P.O. Box 942732
400 P Street
Sacramento, CA 94234-7320
(916) 324-2443

Commonwealth of the Northern Mariana Islands

Department of Public Health and Environmental Services
Division of Environmental Services
Commonwealth of the Northern Mariana Islands
Saipan, CM 96510

Guam

Environmental Protection Agency
P.O. Box 2999
Agana, GU 96910
(671) 646-8863/5

Hawaii

Department of Health
Environmental Protection and Health Services Division
P.O. Box 3378
Honolulu, HI 96801
(808) 548-4139

Nevada

Division of Environmental Protection
Waste Management Section
Room 221
201 South Fall Street
Carson City, NV 89710
(702) 885-4670

Trust Territory of the Pacific Islands

Environmental Protection Board
Trust Territory of the Pacific Islands
Office of the High Commissioner
Commonwealth of the Northern Mariana Islands
Saipan, CM 96950

Region 10

Alaska

Department of Environmental Conservation
Division of Environment
Juneau, AK 99811-1800
(907) 465-2666

Idaho

Department of Health and Welfare
Division of Environment
Bureau of Hazardous Materials
450 West State Street
Boise, ID 83720
(208) 334-5897

Oregon

Department of Environmental Quality
Hazardous and Solid Waste Division
811 S.W. 6th Avenue
Portland, OR 97204
(503) 229-5356

Washington

Department of Ecology
Office of Land Programs
Hazardous Waste Cleanup Program
Mail Stop PV-11
Olympia, WA 98504-8711
(206) 438-3039

Appendix B

State Agencies for Obtaining Well Construction Record Information

J. Douglas Cook
Dept. of Environmental Mgmt.
Water Supply Program
1751 Federal Dr.
Montgomery, AL 35130

Dept. of Water Resources
Water Resources Specialist
99 E. Virginia
Phoenix, AZ 85004
602-255-1581

Edwin A. Ritchie
Dept. of Water Resources
Div. of Planning
Box 388
Sacramento, Ca 95802
916-445-7314

Attn: Well Drlg. Board
Dept. of Consumer Protection
Div. of Licensing/Protection
165 Capitol Av. Rm. G7A
Hartford, Ct. 06106
203-566-3275

Central Contact:
Dept. of Environmental Reg.
Ground Water Section
2600 Blairstone Road
Tallahassee, FL 32399

Dept. of Environmtal. Conserv.
Drinking Water Prg.
Pouch O
Kimeai, AL 99811
907-465-2653

Kenneth T. Acklin
Water Well Const. Commission
2915 S. Pine St.
Little Rock, AR 72204
501-666-8379

Robert Longenbaugh
Division of Water Resources
1313 Sherman Street, Rm. 818
Denver, CO 80203
303-866-3581

Philip Cherry
Dept. of Natural Resources
Div. of Environmental Control
P.O. Box 1401
89 Kings Highway
Dover, DE 19903
302-736-4793

Water Well Stds. Adv. Council
19 Martin Luther King Jr. Dr. SW
Atlanta, GA 30334
404-656-3214

Div. of Water & Land Develop.
P.O. Box 373
Honolulu, HI 96809
808-548-7539

James Mills
Dept. of Health
525 W. Jefferson
Springfield, IL 62761
217-782-5830

Darrell McAllister
Dept. of Natural Resources
Wallace State Office Bldg.
Des Moines, IA 50319
515-281-8869

David Trimble
Division of Water
Ground Water Section
18 reilly Road
Frankfort, KY 40601
502-564-3410

John Beal
Dept. of Water Resources
Statehouse
Boise, ID 83720
208-334-7956

Karyl Schmidt
Dept. of Env. Mgmt.
105 S. Meridian St.
Indianapolis, IN 46206
317-232-8713

Darrel Plummer
Dept. of Health & Environment
Bldg. 740, Forbes Field
Topeka, KS 66620
913-296-1500

Bo Bolourchi
Dept. of Transportation & Dev.
Water Resources Section
P.O. Box 94245
Baton Rouge, LA 70804
504-379-1434

Dept. Health & Mental Hygiene
201 Preston Street
Baltimore, MD 21201
301-225-5850

Dept. of Environmental Quality
Engineering
One Winter Street
Boston, MA 02108
617-292-5654

James Nye
MN Dept. of Health
717 Delaware Street SE
Minneapolis, MN 55440
612-623-5339

Dept. of Natural Resources
Geological & Land Survey Div.
P.O. Box 250
111 Fairgrounds Road
Rolla, MO 65401
314-364-1752

Dept. of Water Resources
Box 94676
Lincoln, NE 68509
402-471-2363

Rick Scofield
NH Water Well Board
Box 2008
Concord, NH 03301
603-271-3406

State Engineer's Office
Div. of Water Rights
Battan Memorial Bldg.
Santa Fe, NM 87501
505-827-6120

Don Keech
Dept. of Public Health
P.O. Box 30035
Lansing, MI 48909
517-335-8300

Dept. of Natural Resources
Bureau of Land and Water
P.O. Box 10631
Jackson, MS 39209
601-961-5200

Diana Cutler
Board of Water Well Contractor
Nat. Resources & Conserv. Dept
1520 E. Sixth Avenue
Helena, MT 59620
406-444-6643

Office of State Engineer
Div. of Water Resources
201 S. Fall St.
Carson City, NV 89701
702-885-4380

Gail Carter
NJ Geological Survey
CN-029
Trenton, NJ 08625
609-984-6587

Dept. of Envir. Conservation
Region 1 - Water Supply Prg.
Bldg. 40, SUNY
Stony Brook, NY 11794

Dept. Nat. Res. & Comm. Dev.
Div. of Environmental Mgmt.
P.O. Box 27687
Raleigh, NC 27611
919-733-5083

Dept. of Natural Resources
Div. of Water
Foundatin Square
Columbus, OH 43224
614-265-6744

Water Resources Dept.
555 - 13th St. NE
Salem, OR 97310
503-378-8456

John Hatch, Chief Engineer
Dept. of Water & Nat. Resources
Water Rights Div.
Joe Foss Bldg.
523 E. Capitol
Pierre, SD 57501
605-773-3352

Rick Nelson
Dept of Health
1200 Missouri Av.
Mismarck, ND 58501
701-224-2354

Dept. of Environmental Res.
Topographic & Geologica Survey
P.O. Box 2357
Harrisburgh, PA 17120
717-787-5828

Private Water Supply Section
Dept. of Natural Resources
Box 7921
Madison, WI 53707
608-267-7649

Dept. of Health & Env. Control
Ground Water Protection Div.
2600 Bull Street
Columbia, SC 29201
803-734-5331

Ground Water Protection Div.
150 9th Avenue N
Terra 7th Floor
Nashville, TN 37203
615-741-0690

Texas Water Commission
P.O. Box 13087
Austin, TX 78711
512-463-7998

Dept. of Env. Mgmt.
83 Park St.
Providence, RI 02903

James Ashley
Dept. of Env. Conservation
103 S. main St.
Waterbury, VT 05676
802-244-5638

William Miller
Dept. of Ecology
Mail Stop PV-11
Olympia, WA 98504
206-459-6045

OK Water Resources Board
P.O. Box 53585
Oklahoma City, OK 73152
405-271-2555

Jerry Bronicel
State Engineer's Office
1636 W. North Temple, Rm. 205
Salt Lake City, UT 84116
801-533-6071

Terry Wagner
State Water Control Bd.
2111 N. Hamilton St.
Richmond, VA 23230
804-367-6406

Chet Fleming
Dept. of Health
1800 Washington St. East
Charleston, WV 25305
304-348-2981

Richard Stockdale
State Engineer's Office
Herschler Bldg. 4th Floor
Cheyenne, WY 82002
307-777-7354

Appendix C

Statements of Meaning

The following statements of meaning are meant to provide some clarification as to possible definitions that will be developed for the elements in the Minimum Set of Data Elements.

LATITUDE

The angular distance north or south from the Earth's equator measured through 90 degrees. The length of a degree varies from 68.704 statute miles at the equator to 69.407 at the poles because of the flattened configuration of the Earth. The length of a second is approximately 100 feet. [USGS]

LONGITUDE

The angular distance, measured in degrees, due east or west from the prime meridian that runs between the north and south poles and passes through Greenwich, England. The length of a degree varies from 69.95 statute miles at the Equator to zero miles at the poles. The length of a second is a little over 100 feet at the equator and about 78 feet at the 40 degree latitudinal parallel which passes through the approximate middle of the United States. [USGS]

METHOD OF MEASURE FOR LATITUDE/LONGITUDE

The method used to determine the latitude/longitude such as surveyed, from a USGS quad sheet, and so forth.

SOURCE AGENCY FOR LATITUDE/LONGITUDE DATA

The Agency that reported the data.

STATE FEDERAL INFORMATION PROCESSING STANDARD CODE (FIPS)

An established standard that is a two digit number representing the state in which the well or spring is physically located.

COUNTY FEDERAL INFORMATION PROCESSING STANDARD CODE (FIPS)

An established standard that is a three digit number representing the county or county equivalent in which the well or spring is physically located.

ALTITUDE

The altitude of the land surface at the well or spring above or below mean sea level, in feet, National Geodetic Vertical Datum.

WELL/SPRING AND FACILITY IDENTIFICATION

An identification to be developed for each well or spring, that never changes and is never duplicated.

An identification, where applicable, that establishes a tie or linkage between a well or a spring and the facility on which it is located.

USE OF WELL

The principal use of a well or spring or the purpose for which the well was constructed (the former always holds precedence over the latter). [after USGS]

DEPTH OF WELL AT COMPLETION

The elevation at the completion depth of the well in feet above or below mean sea level (NGVD).

DEPTH TO TOP OF OPEN INTERVAL

Depth to the point where the opening begins, in feet below land surface. The first section of the opening always begins at depth 0. [after USGS]

DEPTH TO BOTTOM OF OPEN INTERVAL

Depth to the bottom of the open interval, in feet below land surface. [after USGS]

LOCATION OF WELL LOG

The physical location of the well log, such as the Agency name and address where the log is located.

TYPE OF WELL LOG

Identifies the type of well log: a physical description of the rock cuttings of the different formations penetrated, such as a driller's log; or a continuous recording of the electrical, radioactive, acoustic and/or other properties of the penetrated formations, such as an electric log.

SOURCE AGENCY FOR SAMPLE DATA

The Agency that reported the data.

SAMPLE DATE

The date on which the sampling event occurred.

SAMPLE IDENTIFICATION

An identification to be developed to uniquely identify each sample taken, it may include several factors, such as sampling purpose, field conditions, field protocol.

PARAMETER MEASURED

The specific chemical for which an analysis is made.

CONCENTRATION/VALUE

The numerical value (concentration or quantity) detected by the parameter test (in standard units).

CONFIDENCE FACTOR

An element that can include field and laboratory quality assurance, and other factors that provide the degree of confidence the data source has in the value reported.

DEPTH TO WATER

The water level at the well or spring, in feet below land surface. [after USGS]

MEASUREMENT QUANTIFICATION

A method of quantification of a parameter.

From EPA Workshop to Recommend A Minimum Set of Data Elements for Ground Water. Workshop Findings Report, EPA 440/6-88-005, U.S. Government Printing Office, Washington, D.C. (1988).

Appendix D

Software Tools

C-Map — C-Map Computer Mapping, Display and Analysis, User Manuals v. 1.07. Center for Remote Sensing, Michigan State University, East Lansing, MI 48824-111. 517-353-7195.

Statgraphics — Statgraphics v. 3.0, STSC, Inc. 2115 East Jefferson St., Rockville, MD 20852. 301-984-5000.

Geobase4 — Cross-Sections and Chemistry by Hall Groundwater Consultants. Earthware of California, 30100 Town Center Drive, Suite 0196, Laguna Niguel, CA 92677. 714-495-5727.

Surfer — Surfer, Golden Software, Inc. P.O. Box 281, Golden, CO 80420. 303-279-1021.

Foxbase+ — Fox Software, Inc. 27493 Holiday Lane, Perrysburg, OH 43551.

WELLKEY — Michigan Department of Natural Resources, Land and Water Management Division (1989), Statewide Groundwater Database. 517-334-6912.

AQUIPRO — Aquifer vulnerability software. Dr. Richard Passero, Institute of Water Sciences, Western Michigan University, Kalamazoo, MI 49008. 616-387-5505.

CORE CON — File or point conversion software for Michigan to convert Lat-long coordinates to State Plane. Michigan Department of Natural Resources, 1989.

Plane-PC Interactive Grid Coordinate Conversion for Personal Computers. National Mapping Program Technical Instructions, Department of the Interior, USGS National Mapping Division, 1984.

Other Tools

Loran-C — MicroLogic Inc. 9610 Desoto Ave., Chatsworth, CA 91311. 818-998-1216. Geo-location device.

Global Positioning Systems (GPSs) — Magellan by Magellan Systems Corp. 260 E. Huntington Dr., Monrovia, CA 91016. 818-358-2363. GPS Pathfinder from Trimble Navigation 645 North Mary Ave., Sunnydale, CA 94008-3642. 408-730-2906.
Geolink by Geo Research, Inc. 2815 Montana Ave. Billings, MT 59101. 1-800-829-4474

Poly-Pram — On-site water tester for conductivity, pH, temperature, from GISCO Geophysical Instrumentation and Supplies Co., 4665 Joliet St., Denver, CO 80329. 303-371-1940, or from Presto-Tech, 7321 N. Figueroa, Los Angeles, CA 90041. 213-257-7585.

Appendix E

Common GIS Functions

The following list of GIS functions and their definitions is from a compilation produced by a federal agency GIS coordinating committee. It is one list of definitions, but there are others. The definitions of GIS functions have not been standardized, and there is a good deal of variation in them. For comparison and an expansion of the list, see the GIS functions in the GIS Survey Report following this list.

1. BASIC SYSTEM CAPABILITIES

Digitizing — Digitizing is the process of converting point and line data from source documents to a machine-readable format.

Edgematching — Edgematching is the process of joining lines and polygons across map boundaries in the creation of a "seamless" database. The join should be topological as well as graphic; that is, a polygon so joined should become a single polygon in the database and a line so joined should become a single line segment.

Polygonization — Polygonizing is the process of connecting arcs ("spaghetti") to form polygons.

Labeling — This process transfers labels describing the contents (attributes) of polygons and the characteristics of lines and points to the digital system. This input of labels must not be confused with the process of symbolizing and labeling output below.

Reformatting digital data for input from other systems — Data previously digitized are made accessible through an interface or converted by software to the system format, and made to be topologically useful as well as graphically compatible.

Reformatting for output to other systems — This function is the inverse of the previous one. Internal data are reformatted to meet the requirements of other systems or standards.

Database creation and management — Data are typically digitized from mapsheets, and may be edgematched. The creation of a true "seamless" database requires the establishment of a map sheet directory and may include tiling to partition the database.

Raster/vector conversion — The ability to convert data between vector and raster forms with grid cell size, position, and orientation selected by the user.

Edit and display on input — This function allows continuous display and editing of input data, usually in connection with digitizing.

Edit and display on output — The ability to preview and edit displays before creation of hard copy maps.

All symbolizing — To create high-quality output from a GIS, it is necessary to be able to generate a wide variety of symbols to replace the primitive point, line, and area objects stored in the database.

Plotting — Creation of hard copy output.

Updating — Updating of the digital database with new points, lines, polygons, and attributes.

Browsing — Browse is used to search the database to answer simple locational queries and included pan and zoom.

2. DATA MANIPULATION AND ANALYSIS FUNCTIONS

Create lists and reports — This is the ability to create lists and reports on objects and their attributes in user-defined formats and to include totals and subtotals.

Reclassify attributes — Reclassification is the change in value of a set of existing attributes based on a set of user-specified rules.

Dissolve lines and merge attributes — Boundaries between adjacent polygons with identical attributes are dissolved to form larger polygons.

Line thinning and weeding — This process is used to reduce the number of points defining a line or set of lines to a user-defined tolerance.

Line smoothing — Automatically smooth lines to a user-defined tolerance, creating a new set of points.

Complex generalization — Generalization that may require change in the type of an object or relocation in response to cartographic rules.

Windowing — The ability to clip features in the database to some defined polygon.

Centroid calculation and sequential numbering — calculate a contained, representative point in a polygon and assign a unique number to the new object.

Spot heights — Given a digital elevation model, interpolate the height at any point.

Heights along streams — Given a digital elevation model and a hydrology net, interpolate points along streams at fixed increments of height.

Contours (isolines) — Given a set of regularly or irregularly spaced point values, interpolate contours at user-specified intervals.

Elevation polygons — Given a digital elevation model, interpolate contours of height at user-specified intervals.

Watershed boundaries — Given a digital elevation model and a hydrology net, interpolate the position of the watershed between basins.

Scale change — Perform the operations associated with change of scale, which may include line thinning and generalization.

Rubber sheet stretching — The ability to stretch one map image to fit over another, given common points of known locations.

Distortion elimination — The ability to remove various types of systematic distortion generated by different input methods.

Projection change — The ability to transform maps from one map projection to another.

Generate points — The ability to generate points and insert them in the database.

Generate lines — The ability to generate lines and insert them in the database.

Generate polygons — The ability to generate polygons and insert them in the database.

Generate circles — The ability to generate circles defined by the center point and radius.
Generate grid cell nets — The ability to generate a network of grid cells given a point of origin, grid cell dimension, and orientation.

Generate latitude/longitude nets — The ability to generate graticules for a variety of map projections.

Generate corridors — This process generates corridors of given width around existing points, lines, or areas.

Generate graphs — Create a graph illustrating attribute data by symbols, bars, or fitted trend line.

Generate viewshed maps — Given a digital elevation model and the locations of one or more viewpoints, generate polygons enclosing the area visible from at least one viewpoint.

Generate perspective views — From a digital elevation model, generate a three-dimensional block diagram.

Generate cross-sections — Given a digital elevation model, show the cross-section along a user-specified line.

Search by attribute — The ability to search the database for objects with certain attributes.

Search by region — The ability to search the database within any region defined to the system.

Suppress — The ability to exclude objects by attribute (the converse of selecting by attribute).

Measure number of items — The ability to count the number of objects in a class.

Measure distances along straight and convoluted lines — The ability to measure distances along a prescribed line.

Measure length of perimeter of areas — The ability to measure the perimeter of a polygon.

Measure size of areas — The ability to measure the area of a polygon.

Measure volume — The ability to compute the volume under a digital representation of a surface.

Calculate-arithmetic — The ability to perform arithmetic, algebraic and Boolean calculations separately and in combination.

Calculate bearings between points — The ability to calculate the bearing (with respect to true north) from a given point to another point.

Calculate vertical distance or height — Given a digital elevation model, calculate the vertical distance (height) between two points.

Calculate slopes along lines (gradients) — The ability to measure the slope between two points of known height and location or to calculate the gradient between any two points along a convoluted line that contains two or more points of known elevation.

Calculate slopes of areas — Given a digital elevation model and the boundary of a specified region (a part of a watershed), calculate the average slope of the region.

Calculate angles and distances along linear features — Given a prescribed linear feature, generalize its shape into a set of angles and distances from a start point, at user-set angular increments, and constrained to any known points along the linear feature.

Subdivide area according to a set of rules — Given the corner points of a rectangular area, topologically subdivide the area into four quarters.

Locations from traverses — Given a direction (one of eight radial directions) and distance from a given point, calculate the end point of the traverse.

Statistical functions — The ability to carry out simple statistical analyses and tests on the database.

Graphic overlay — The ability to superimpose graphically one map on another and display the result on the screen or on a plot.

Point in polygon — The ability to superimpose a set of points on a set of polygons and determine which polygons (if any) contain each point.

Line in polygon — The ability to superimpose lines on a set of polygons and determine which polygons (if any) contain each line.

Line on polygon overlay — The ability to superimpose a set of lines on a set of polygons, breaking the lines at intersections with polygon boundaries.

Polygon overlay — The ability to overlay digitally one set of polygons on another and form a topological intersection of the two.

Sliver polygon removal — The ability to delete automatically the small sliver polygons that result from a polygon overlay operation when certain polygon lines on the two maps represent different versions of the same physical line.

Line of sight — The ability to determine the intervisibility of two points, or to determine those parts of pairs of lines or polygons which are intervisible.

Nearest neighbor search — The ability to identify points, lines, or polygons that are nearest to points, lines, or polygons specified by location or attribute.

Shortest route — The ability to determine the shortest or minimum cost route between two points or specified sets of points.

Contiguity analysis — The ability to identify areas that have a common boundary or node.
Connectivity analysis — The ability to identify areas or points that are (or are not) connected to other areas or points by linear features.

Complex correlation — The ability to compare maps representing different time periods, extracting differences, or computing indices of change.

Weighted modeling — The ability to assign weighting factors to individual data sets according to a set of rules and to overlay those data sets and carry out reclassify, dissolve, and merge operations on the resulting concatenated data set.

Scene generation — The ability to simulate an image of the appearance of an area from map data. The image would normally consist of an oblique view, with perspective.

Network analysis — Simple forms of network analysis are covered in shortest route and connectivity. More complex analyses are frequently carried out on network data by electrical and gas utilities, communications companies, etc. These include the simulation of flows in complex networks, load balancing in electrical distribution, traffic analysis, and computation of pressure loss in gas pipes. In many cases, these capabilities can be found in existing packages that can be interfaced to the GIS database.

Table 1. General System Characteristics

GIS System	Organization	First Installed	Number Installed	Number of Users	Pricing	Source Code Avail.	Source Code Price	Custom ization ?	User Support Fee
A-BASE	Brighter Images, Inc.	1989	70	80	$1,000			●	
ACDS	ACDS Graphic Systems Inc.	1985	110	225	$12,000			●	15 percent
ADDMAPS	Cobham Digital Services Ltd.	1989	3		$10,000			●	10 percent
ARC/INFO	Environmental Systems Research Institute, Inc. (ESRI)	1982	4000+	10,000+	NR			●	NR
ATLAS*DRAW	Strategic Mapping, Inc.	1988			$750				
ATLAS*GIS	Strategic Mapping, Inc.	1990			$2,495				
ATLAS*GRAPHICS	Strategic Mapping, Inc.	1988			$495				
ATLAS*MAPMAKER	Strategic Mapping, Inc.	1986			$395				
ATLAS/AGIS	Delta Data Systems, Inc.	1987	15	30	$6,995			●	$2,000
ATOM	GeoSpectra Corporation	1983	6		$200,000			●	
AUDRE	AUDRE, Incorporated	1983	50+	100+	$50,000			●	NR
Autoroute Plus	Nextbase Ltd	1989	5000	5000	£299				
CableCad	Enghouse Systems Ltd	1986	700		$12,000			●	12 percent
CADCore/Hybrid	Hitachi America, Ltd.	1989			$3,600			●	
CADKEY	CADKEY, Inc.	1985	55000+		$3,195			●	$695
Cartographer	Mercator Software Systems	1990	7	30	$995			●	Free telephone support
CDM: Geo. Info. Mngr.	Camp, Dresser & McKee	1990			$25,000			●	$2,500 for 1st Yr.
CDM: Map Manager	Camp, Dresser & McKee	1986	50+	100+	$15,000			●	$1,500 for 1st Yr.
Census Windows:Tiger Tools	Geovision, Inc.	1989	3	25	$1,495			●	
CPS-3	Radian Corp					●		●	
CRIES-GIS	Michigan State University	1980	52	200-300	$1,500			●	
DIGI-CAD	Air-Land Surveys, Inc.	1988	1						
EARTH ONE	C. H. Guernsey				$12,000			●	$2,070
EARTH/INFO	Earth Info Sciences	New			NR			●	By application
EASI/PACE	PCI Remote Sensing Corp.	1980	275	400	$3,000 PC, $7,000	●	$7,000	●	$2,400
EIFM	Geographic Systems Corporation	1981	6		NR	●	NR	●	NR
ERDAS	ERDAS, Inc.	1979	1200	5000	$2,000+			●	$2,000 - $3,000
FMS/AC	Facility Mapping Systems, Inc.		650+	650+	$3,000+	●	$1,000	●	12 % or $300
GDS	McDonnell Douglas	1980	1000+	4000+	$9,500			●	12 percent
Genamap	Genasys II, Inc.	1986	120+	300+	NR	●	NR	●	NR

NR = Not Reported

	Data Structures									
Computers	Raster	Topological Vector	Non-topological Vector	Quadtree	TIN	Other	Geographic Reference (Lat/Long)	Map Projections Conversion	Includes Map Digitizing Package	DBMS Interfaces
DOS , OS/2	●	●	●		●					Dbase III+
IBM RS/6000, RT, PS/2 - AIX; OSF & UNIX International - UNIX		●	●	●			●	●	●	Oracle
PC Compatibles - DOS		●						●	●	Dbase IV, DATAFLEX, DATAEASE
Apollo, DEC, DG, HP, IBM, Intergraph, Prime, SUN, VAX Tektronix (see page ???)	●	●	●	●	●	Linkages to CADD, DBMS, scanned documents	●	●	●	INFO, Oracle, INGRES, Sybase, Informix, DB2, Rdb, SQL, DS, Dbase (3,4)
PC XT, AT, PS/2 - DOS			●				●	●	●	
PC XT, AT, PS/2 - DOS			●				●	●	●	Dbase III and compatible
PC XT, AT, PS/2 - DOS			●							
Macintosh			●				●	●	●	
PC AT compatibles	●	●	●				●	●		Dbase III
VAX - VMS; Alliant Concentrix; Sun - UNIX	●						●	●	●	
Apollo; PC Workstations	●						●	●	●	Ingress, Prolog, Oracle, Informix
PC, PS/2 - DOS	●	●	●				●			Dbase III+ or IV or comp.
DOS, OS/2 EE, AIX		●	●				●		●	Proprietary
DOS	●	●							●	Oracle
PC XT, AT - DOS	●						●			
Macintosh		●					●			Mac database w/ cut/copy/ paste & TEXT data input
PC AT - DOS, UNIX	●	●	●				●	●	●	PROGRESS
PC AT - DOS			●				●		●	PROGRESS
80286/80386 - DOS 3.3+, Windows, (MSCDEX for GEOdisc CD ROM optional)		●				System is GEO-processing tool for Tiger	●	●		VIA DDE, Excel Gate, Windows-Based DBMS
Most UNIX, SGI, Apollo, HP, VAX, IBM, Interpro, Sun; 386 - DOS		●	●		●	x,y,z grid format	●	●	●	Oracle, INFO, EMPRESS, DB2, Informix, Sybase, Aegis, most SQL
DOS compatible systems	●	●					●		●	Dbase III
		●			●		●			
DOS compatible (requires math co-processor)	●	●	●				●		●	Proprietary
PS/2 or compatible - DOS		●					●	●	●	ENABLE/Dbase
PC - DOS; SUN - UNIX; Silicon Graphics - UNIX; Stardent ST-3000 - UNIX; VAX - VMS; CYBER - NOS/VE	●		●			N-bit Raster, ASCII, VOXEL	●	●	●	
IBM 9370, 43XX, 30XX - MVS, IMS, VM, DB2, SQL 5080; PS/2			●				●	●	●	DB2, SQL, IMS
386 PC - DOS; Sun 386i; Sun 3 series; SPARCstation, Microvax - UNIX; VAX - VMS	●						●	●	●	Systems supported through ERDAS-ARC/INFO Live Link
PC - DOS; SUN 386i; SPARC (June'90); Dec Station 3100		●	●				●	●	●	Dbase, Fox (Oracle fall '90)
Digital VAX - VMS; VAXstations - VMS; DECstations - ULTRIX	●	●	●	●	●		●	●	●	Any SQL based database
IBM RS/6000; HP9000 300/800; SUN SPARC; Apollo; Silicon Graphics; Concurrent; WANG; 386-based PCs	●	●	●		●	R-trees, B-trees	●	●	●	INGRES, Oracle, INFORMIX, HP ALLBASE, SQL 400, DB2

Table 1. General System Characteristics (continued)

GIS System	Organization	First Installed	Number Installed	Number of Users	Pricing	Source Code Avail.	Source Code Price	Custom-ization ?	User Support
Geo/SQL	Generation 5 Technology, Inc.	1988	350	550	$9,500			●	10 percent
Geographic Facilities Information System	IBM Corporation	1977	200+	200+	variable			●	
Geographic Road Network Data Base (GRNDB)	Division of Government Research	1982	2	20	NR	●	NR	●	NR
GIFM	Geographic Systems Corporation	1981	4		NR	●		●	NR
GIS CAD	Daniel Geophysical	1989	10	20	$4,500			●	1.5 percent
GIS Plus	Caliper Corporation	1987			$2,995			●	$450
GIS/AMS	GeoVision Systems, Inc.	1976	80	300	$100,000			●	$15,000
GRASS	U. S. Army COE, Construction Engineering Engineering Research Laboratory	1985	1000+	1000+	NR	●	included	●	NR
GWN-GIS	GWN Systems Inc.	1989	10	10	$6,000			●	$600
Chains	Hunter GIS Inc.	1987	25		$5,000 - 15,000			●	10 - 15 percent
IDRISI	Clark Graduate School of Geography	1987	1500+		$200				$50
INFO-MAP	Claymore Services Ltd	1988	115	180	£995			●	£150 after 1st year
INFORMAP	Synercom Technology, Inc.	1974						●	
InSite*USA	CACI Marketing Systems	1988	40	150	4			●	12.5% of price
KDMS	Kork Systems, Inc.	1981	100	100	$4,900			●	$1,225
LAND	Geographic Systems Corporation	1986	4		NR	●	NR	●	NR
LandTrak	GeoBased Systems	1983	300	250	$3,800			●	$2,500
MacGIS	University of Oregon	1989	300	500	$300			●	
MacGIS	CLEARS - Cornell University	1988	100+	100+	$150				
MACMAP	KLIK	1989	150	200	$5000-$6000			●	12 percent
Manatron	Manatron, Inc.	1980	60+	75+	$15,000			●	15% of software license
Map and Image Processing System (MIPS)	MicroImages	1987			$2,000 - 5,000			●	
MapBox/Dragon	Decision Images, Inc.	1988	150+	150+	$995			●	
MapGrafix	ComGrafix, Inc.	1987	200	250	$8,500			●	$500
MAPII: Map Processor	John Wiley & Sons	1989	300	3500	$95.95				
MapInfo	MapInfo Corp.	1987	10,000+	10,000+	$995			●	$145
MapPower	Map Power Corporation	Beta			$17,000			●	10%
MARS (Mapping Applications Resource System)	Montage Information Systems, Inc.	1990	2	4	$1,495			●	
MGE	Intergraph Corporation	1989	486		$8,300			●	included in maint. contract
MIMS	Computer Mapping Company	1988	6	6	$4,200			●	$475
MOSS	Autometric, Inc.	1978			Public Domain	●	Duplication	●	$5,000
Nucor	Nucor Computing Resources Inc.	1988	47	188	$4,000			●	12 percent
PAMAP GIS	Pamap Graphics Ltd.	1986	200	300	$7,500+	●	NR	●	20 percent

NR = Not Reported

Computers	Data Structures: Raster	Topological Vector	Non-topological Vector	Quadtree	TIN	Other	Geographic Reference (Lat Long)	Map Projections Conversion	Includes Map Digitizing Package	DBMS Interfaces
386 PC - DOS; Sun SPARCstation - UNIX: OS/2 (Q4 1990)		●	●				●	●	●	ANSI SQL 89 compliant; RBase; Oracle, others
IBM 9370 midrange 3090 high end 370 base architecture, VM or MVS, 5080 display to PS/2			●			Hierarchical	●	●	●	Geo-facilities DataBase Mgt. (GDBS)-IMS, Geo-Manager-DB28, SQL DS
IBM - MVS; (Under development - DEC-VMS & DEC-ULTRIX)						Roadway Inventory Structured	●	●		Statistical Analysis System (SAS)
IBM 9370, 43XX, 30XX, MVS, IMS, VM, DB2, SQL 5080, PS/2			●				●	●	●	DB2, SQL, IMS
386 PC - DOS; various UNIX			●				●	●	●	RBase for DOS, various DBMSs for UNIX
80286, 80386, 80486 PC - DOS 2.0 or higher		●					●	●	●	1-2-3; Generic with ASCII export capability
DEC VAX - VMS, ULTRIX; SUN SPARC - UNIX; IBM RS/6000 - AIX	●	●		●	●		●	●	●	Oracle - embedded. Copy included with software.
Sun; DECstations; Intergraph; Tektronix; MASSCOMP; Macintosh II; PC 386, 486 - UNI; DG; IBM PS/2, 600	●	●					●	●	●	INFORMIX, RIM
PC - DOS		●	●						●	Dbase
286/386/486 - DOS w/15Mb 6 RAM	●	●	●						●	Micro RIMIS RBase
PC/XT/AT, 386, PS/2 - DOS	●		●				●	●	●	Dbase III, Professional File
PC EGA/VGA Colour, 640k RAM - DOS	●						●		●	Any via ASCII, DXF files
Digital VAX - VMS, ULTRIX; Tektronix IBM PC; NCD X Terminal; Mac; DG Aviion - UNIX; Sun SPARCstation; Apollo Aegis	●	●	●				●	●	●	Oracle and others
PC 386 (w/387) 640K - DOS, 320MB Hard Drive			●				●		●	Dbase to interface with many others
386 PC's - DOS; VAX - VMS 4.7+; PDP TSX; RT-11			●					●	●	
IBM 9370, 43XX, 30XX - MVS, IMS, VM, DB2, SQL 5080; PS/2		●	●				●	●		IMS, DB2, SQL, DS
PC - DOS		●					●	●	●	Proprietary database
Macintosh	●						●		●	Hypercard
Macintosh	●		●						●	Any Mac DBMS
Macintosh			●			Object Database			●	Exel, UD, Fox base Ornis
Unisys B20 family, BTOS, (DOS by late summer)			●				●	●	●	Fasport
80386 & 80286 w/ AT BUS; PS2 w/ Microchannel, Windows 386	●	●	●				●	●	●	Dbase III, Dbase IV, INFO, RBASE
PC or PS/2 - DOS, graphics adapter	●									
Macintosh			●				●	●	●	ASCII text bidirectional exchange
Macintosh	●							●		
PC - DOS			●			Proprietary	●		●	Dbase, FoxBase, ASCII
Macintosh	●		●				●	●	●	With CL1 & SQL links all applicable systems
80386/80286 PC's - DOS, WINDOWS 2Mb RAM			●				●		●	Virtually any PC based Soft, DBF, etc., Dbase, Q&E, Superbase, Excel
Intergraph Unix workstations	●	●	●		●		●	●	●	ORACLE, INGRES, INFORMIX, DB2
PC - DOS						Object-oriented (vector)	●	●	●	Dbase III & compatible
Prime - Primos; DG - AOS/VS,; Tektronix 4300 & XD88 - UTEK	●	●	●				●	●	●	Internal, Oracle, DG/SQL, ASCII
AT - DOS	●		●				●	●	●	Zim
PC/AT, PS/2 - DOS, OS/2; SUN 4, SPARCstation - UNIX; INTERPRO/ CLIPPER - UNIX; DG Aviion - UNIX; DEC VAX - VMS	●	●	●				●	●	●	ORACLE, RDB, INGRES, Dbase, RBase & others

Table 1. General System Characteristics (continued)

GIS System	Organization	First Installed	Number Installed	Number of Users	Pricing	Source Code Avail.	Source Code Price	Custom-ization ?	User Support
pMAP	Spatial Information Systems, Inc.	1986	150	200	$895				
QUIKMap	ESL Environmental Services Ltd.	1989	90	100	$800			●	NR
SPANS	TYDAC Technologies Corp.	1985	1000	1200	$6,000			●	12 percent
STRINGS	GeoBased Systems	1979	300	200	$9,000			●	$2,100
Survey System	C & G Software Systems, Inc.	1979	900	2500	$2,300				$400+
System 600	International Imaging Systems	1984	300	1,000	$15,000	●	included	●	18 percent
System 9	Computervision	1987	50	200	$12,500 - 20,000			●	15 percent
Terra-Mar GIS	Terra-Mar Resource Info. Services, Inc.	1983			$2,495				12 percent
Terrasoft	Digital Resource Systems (Canada) Ltd	1985	200	200+	$10,500			●	●
TIGRIS	Intergraph Corporation	1988			$10,000			●	Included in maint. contract
Topologic	Geometria Software Laboratories Co.	1987	22	60 - 70	$2,000 - 7,000			●	One year free
UltiMap	UltiMap Corporation	1983	56					●	
USEMAP 4.2	Waldeck Geosoftware	1989	10	40	$1,500 - $3,000			●	10 percent
Vortext	Aangstrom Precision Corporation	1988	16	60+	$19,500			●	15% includes maintenance
WIFM	Geographic Systems Corporation	1989	1		NR	●	NR	●	NR
Windows/on the World	Geovision, Inc.	1986	1200	1200	$595	●	NR	●	NR
XMap	DeLorme Mapping Company	1988			$4995			●	15 percent
No System Name	Atlantic Aerial Surveys, Ltd.								
No System Name	CADworks Inc.								
No System Name	DIPIX Technologies Inc.	1978		200+				●	
No System Name	GTX Corporation								
No System Name	Land Innovation, Inc.							●	
No System Name	Leica USA, Inc. (formerly Wild LEITZ USA)								
No System Name	Spatial Information Sciences, Inc.								
No System Name	Synergis Technologies, Inc.	1986	17	40	NR			●	
No System Name	VITec - Visual Information Technologies, Inc.								

NR = Not Reported

Computers	Data Structures: Raster	Topological Vector	Non-topological Vector	Quadtree	TIN	Other	Geographic Reference (Lat/Long)	Map Projections Conversion	Includes Map Digitizing Package	DBMS Interfaces
PC - DOS, EGA	●									
PC XT/AT & PS/2 w/ 640k RAM, mouse & math coprocessor			●				●	●	●	Dbase
PS/2 - DOS; PS/2 - OS/2; IBM RS/6000 - AIX	●	●		●	●	Linear Network	●	●	●	OS/2 DBMS, SQL Interface, Oracle, Dbase IV
PC - DOS		●					●	●	●	Ingres, Sybase, Sharebase
PC 286/386/486 - DOS, Novell Network						CTree Binary or ASCII	●	●	●	Dbase III or Later
DEC - VMS; Sun 386i, 3, 4 - UNIX	●						●	●		INGRES
Computervision (SUN) - UNIX		●	●	●	●	Composite features	●	●	●	Empress, Prime Info.
VMS, UNIX, DOS 386/486, Sun	●		●			B-tree	●	●	●	Dbase and proprietary
IBM - DOS	●	●	●				●	●	●	Dbase III+, IV, RBASE, ORACLE
Intergraph UNIX workstations	●	●	●		●	Objects	●	●	●	
OS/2, PC	●	●		●	●				●	Dbase III Plus, OS/2 SQL
HP/Apollo 3500 and 4500		●					●	●	●	All commercial DBMS
PC/AT - DOS	●		●				●		●	Dbase
DEC, Sun SPARCstations, DG Aviion, Intergraph Clipper, Silicon Graphics - UNIX		●	●	●	●	Binary Trees	●	●	●	Most standard DBMS
IBM 9370, 43XX, 30XX - MVS, IMS, VM, DB2, SQL 5080; PS/2			●				●	●	●	DB2, SQL, IMS
80286/80386 - DOS, Windows, MSCDEX for GEOdisc CD-ROM	●	●					●		●	Windows-Based
UNIX-X WINDOWS, SUN, APOLLO, 386IX, X11.3R, UNS-DEC, MS DOS - WINDOWS 2.1, 3.0						Proprietary	●	●		Sybase, MS DOS, DDE
	●	●	●		●		●	●	●	
										Any system that can import CSV files
VAX, MICROVAX, VMS	●						●	●	●	
	●	●								Oracle 5.1b - 6.0
					●		●		●	Built into GIS software
			●							Flat file output only
	●	●	●		●		●	●	●	
IBM - DOS; Sun Workstation - UNIX; DECstation - ULTRIX	●	●	●		●		●	●	●	Various RDBMS, Oracle, INFORMIX, Dbase III
	●						●	●	●	

Table 2. System Functional Capabilities

System Name	Input Data Formats																						
	ARC	AVHRR	DEM	DIME	DLG	DTED	DXF	ELAS	EPSF	ERDAS	ETAK	GIRAS	HPGL	IGES	ISIF	LANDSAT	MOSS	NTF	PICT	SIF	SPOT	TIFF	TIGER	Other
A-BASE			●	●	●		●				●		●									●	●	
ACDS				●	●		●							●	●		●			●				E-File, AMF, IFF
ADDMAPS							●							●										OSTF
ARC/INFO	●		●	●	●	●	●	●		●	●	●	●	●	●	●	●			●	●	●	●	
ATLAS*DRAW																								Proprietary
ATLAS*GIS																								
ATLAS*GRAPHICS																								Proprietary
ATLAS*MAPMAKER																			●					PAINT
ATLAS/AGIS		●	●		●		●	●		●		●				●	●				●		●	TIMS, CZCS
ATOM		●	●			●										●	●				●			
AUDRE																			●			●		
Autoroute Plus																								Dbase III+, DBF
CableCad							●								●					●				IFF
CADCore/Hybrid							●																	
CADKEY																								
Cartographer				●							●												●	
CDM: GIM				●			●																●	
CDM: MM				●			●																●	
Census Windows																							●	
CPS-3							●						●											Any ASCII
CRIES-GIS		●								●						●				●				Intergraph, Arc/Info
DIGI-CAD	●				●		●							●	●					●				
EARTH ONE			●		●		●									●								
EARTH/INFO					●		●																●	ANSII
EASI/PACE	●	●	●		●	●		●		●						●					●	●	●	
EIFM				●	●															●			●	ASCII
ERDAS	●	●	●	●	●	●	●	●							●	●				●	●		●	
FMS/AC				●	●		●				●			●									●	
GDS				●	●		●		●		●			●	●		●	●		●			●	
Genamap			●	●	●	●	●			●	●		●	●	●		●			●			●	GFIS, CALMA, PALETTE, Optigraphics DFAD, ADRG, ITD, TDD, ASCII
Geo/SQL	●		●	●	●		●				●			●									●	
Geographic Facilities Information System				●																			●	IBM , GFIS, IFF
GRNDB	●			●	●																		●	SAS based
GIFM				●	●															●			●	ASCII
GIS CAD			●		●		●																	WHCS, PID, SEGY, TOBIN, GII
GIS Plus					●																●		●	
GIS/AMS				●	●								●	●						●	●	●	●	GINA
GRASS	●	●	●		●	●	●	●								●	●							
GWN-GIS							●																	
Chains					●		●										●							
IDRISI	●	●			●		●	●		●						●					●	●		MAP, Odyssey, CTG

	Measurements				Generate Buffers					Map Algebra Functions					Polygon Operations					
Export Formats	Straight Distance	Curved Distance	Proximity Analysis	Area Measurement	Around Points	Around Polygons	Along Straight Lines	Curved Lines	Weighted	Add/Subtract Maps	Mult./Divide Maps	Exponentiate Maps	Trig. Functions	Differentiate Values	Polygon Overlay	Point in Polygon	Line in Ploygon	Merge Dissolve	Delete Spurious Polygons Overlay	Thiessen Polygon Generation
Dbase III, DXF, DWG	●			●																
DXF, SIF, IGES, E-FILE, IFF	●	●	●	●	●	●	●	●	●	●	●	●	●	●	●	●	●			
DXF, OSTF	●	●		●						●	●		●		●				●	
ARC, DEM, DIME, DLG, DTED, DXF, ELAS, ERDAS, ETAK, GIRAS, IPGL, IGES, ISIF, LANDSAT, MOSS, SIF, SPOT, TIFF, TIGER	●	●	●	●	●	●	●	●	●	●	●	●	●	●	●	●	●	●	●	●
	●			●	●		●													
	●			●	●		●	●							●	●				
PICT, PAINT, TEXT	●																			
ELAS, RASTER, DLGIII, DXF, ASCII	●	●	●	●	●	●	●	●	●	●	●	●	●	●	●	●	●			
										●	●			●						
DXF, SIF, ISIF, HPGL, IGES, DXF, PICT, TIFF, MOSS	●			●					●				●		●	●	●	●	●	
PCX	●														●	●				
DXF, ISIF, SIF, IFF	●	●	●	●																
DXF, ASCII	●	●		●																
DXF, CADL, IGES	●	●	●	●	●	●	●		●						●	●				
PICT, TEXT	●														●	●				
DXF	●	●	●	●	●	●	●	●							●	●	●	●		
DXF	●	●													●					
ARC, DXF, HPGL, TIFF, TIGER, DBF, BIFF, XLS, MICROSTATION										●					●	●	●			
ASCII, DXF, SGF	●			●						●	●	●	●	●		●	●			●
ERDAS-GIS, ASCII			●	●	●	●	●			●	●									
IGES, DXF, SIF, KDMS	●	●		●																
DXF	●	●		●	●	●	●	●							●	●	●			
	●		●	●	●		●													
ARC, SPANS, GINA	●	●	●	●	●	●	●	●		●	●	●	●	●	●					
IFF	●	●																		
ARC/INFO, DLG, DXF, SIF	●	●	●	●	●	●	●	●	●	●	●	●	●	●	●	●	●		●	
DXF, IGES	●	●	●	●	●	●	●	●	●	●	●		●		●				●	●
DIME, DLG, DXF, EPSF, ETAK, IGES, ISIF, MOSS, NTF, SIF, TIGER	●	●	●	●	●	●	●	●	●	●	●	●	●	●	●	●	●	●	●	●
DLG, SIF, IGES, MOSS, DXF	●	●	●	●	●	●	●	●	●	●	●	●	●	●	●	●	●	●	●	●
	●	●	●	●	●	●	●	●	●						●	●	●	●		
IFF	●	●	●	●	●	●	●	●							●	●	●	●	●	
SAS based	●	●	●	●	●	●	●	●	●											
IFF	●	●																		
DXF	●	●		●	●	●	●	●		●	●			●	●			●	●	
HPGL, PCX, CSV	●	●	●	●	●	●	●	●							●	●	●		●	
GINA, IGES	●	●	●	●	●	●	●	●							●	●	●	●	●	
DLG, Digit, DXF, MOSS, ELAS, Generic raster matrix	●	●	●	●	●	●	●	●	●	●	●	●	●	●						
DXF, HPGL	●	●	●	●	●	●	●	●	●	●	●				●	●	●	●	●	
DLG, DXF	●	●	●	●	●	●	●	●		●	●			●	●	●	●	●	●	
ARC, ERDAS, MAP	●		●	●	●	●	●	●	●	●	●	●		●	●	●	●			●

Table 2. System Functional Capabilities (continued)

System Name	Terrain Analysis with DEM Data: Compute Slope Ang.	Determine Compass Aspect	Interpolate Elevation	Line of Sight Viewshed	Generate Elevation Contours	Model Drainage Network	Generate Terrain Cross-sections	Cut/Fill Calculations	Miscellaneous: Anaysis Within Corridor	Boolean Operations-Multiple Maps	Boolean Operations-Multiple Themes	Nearest Neighbor Search	Moving Window Filt.	Comp. Optimal Path	Compute Weighted Optimal Path	COGO Computations	Network Analysis	Vector to Raster Conversion	Raster to Vector Conversion	
A-BASE	●		●		●	●	●	●												
ACDS	●		●		●		●		●	●	●	●	●			●	●			
ADDMAPS	●			●	●		●	●												
ARC/INFO	●	●	●	●	●	●	●	●	●	●	●	●		●	●	●	●	●	●	
ATLAS*DRAW																				
ATLAS*GIS									●											
ATLAS*GRAPHICS																				
ATLAS*MAPMAKER																				
ATLAS/AGIS					●	●	●	●	●	●	●					●		●	●	
ATOM	●	●	●	●	●	●	●	●	●	●	●	●	●			●		●	●	
AUDRE																		●	●	
Autoroute Plus														●	●					
CableCad																●	●			
CADCore/Hybrid																		●	●	
CADKEY																●		●	●	
Cartographer																				
CDM: GIM						●			●	●		●		●	●		●	●	●	
CDM: MM						●			●					●	●		●			
Census Windows																		●		
CPS-3	●		●		●	●	●	●	●	●	●					●				
CRIES-GIS	●	●	●						●	●	●	●	●					●		
DIGI-CAD																				
EARTH ONE	●		●		●				●							●	●	●		
EARTH/INFO									●	●	●	●	●	●			●			
EASI/PACE	●	●	●		●	●	●	●	●	●	●		●			●		●	●	
EIFM																●	●			
ERDAS	●	●	●	●	●				●	●	●	●	●	●	●			●	●	
FMS/AC						●			●	●	●		●			●	●			
GDS	●	●	●	●	●	●	●	●	●	●	●	●	●	●	●	●	●	●		
Genamap	●	●	●	●	●	●	●	●	●	●	●	●	●	●	●	●	●	●	●	
Geo/SQL									●	●	●					●				
Geographic Facilities Information System									●	●	●	●		●	●	●	●			
GRNDB																				
GIFM																●	●			
GIS CAD				●	●		●		●	●	●		●			●			●	
GIS Plus									●					●	●		●			
GIS/AMS				●	●		●		●		●	●		●		●	●			
GRASS	●	●	●	●	●	●	●	●	●	●	●	●	●	●	●		●	●	●	
GWN-GIS	●	●	●	●	●		●	●	●	●	●					●				
Chains	●	●							●	●	●							●	●	
IDRISI	●	●	●	●	●		●		●	●			●	●	●			●	●	

Command Language	Menus	Icons	Multiple Windows	3-D Shaded Relief	3-D Wireframe	3-D Thematic Layer	Single Screen Oper.	Dual Screen Oper.	Multiuser Capability	Customizable Menu	Spatial Query by Cursor Input	Spatial Query by Coordinate Input	X-Windows	SQL	IRDS (FIPS 156)	GKS (FIPS 120)	PHIGS (FIPS 153)	CGM (FIPS 128)	POSIX (FIPS 151-1)	TCP/IP	GOSIP (FIPS 146-1)	ISO Network Stds	Does System Provide On-line Help	Is On-line Help Context-sensitive
User Interface and Display													Support of Standards											
•	•										•	•											No	No
•	•		•				•	•	•	•	•	•	•	•					•	•			Yes	Yes
	•		•	•	•		•	•		•													Yes	No
•	•	•	•	•	•	•	•	•	•	•	•	•	•	•					•	•		•	Yes	Yes
	•						•																Yes	Yes
	•						•	•	•		•	•											Yes	Yes
	•						•																Yes	Yes
	•	•																					Yes	No
	•		•			•		•			•	•											No	No
	•	•	•	•		•	•		•		•	•	•										Yes	No
•	•	•	•				•		•	•	•	•								•		•	Yes	No
	•						•				•	•											No	No
•	•		•					•	•	•	•	•	•						•	•			Yes	Yes
•	•		•				•			•	•	•		•									Yes	No
•	•	•	•	•	•		•	•	•	•	•	•											Yes	Yes
	•	•						•			•												No	No
•	•			•	•	•	•		•	•	•	•	•	•									Yes	Yes
	•						•				•	•		•									Yes	Yes
•	•	•	•				•				•	•											Yes	No
•	•	•	•	•		•	•		•	•	•	•	•	•		•	•						Yes	No
	•				•	•		•			•	•											No	No
																							No	No
•	•			•	•	•	•	•	•		•									•		•	Yes	Yes
•	•	•	•						•	•													Yes	No
•	•	•	•	•		•	•	•	•	•	•	•	•							•			Yes	Yes
•	•	•						•	•	•	•	•		•									Yes	No
•	•		•	•		•	•	•	•	•	•	•	•							•			Yes	Yes
•	•	•	•		•		•	•	•	•	•	•											Yes	No
•	•	•	•	•	•	•	•		•	•	•	•	•	•		•			•			•	Yes	Yes
•	•	•	•	•	•	•	•	•	•	•	•	•	•	•		•		•	•	•		•	Yes	Yes
•	•	•	•	•	•	•	•	•	•	•	•	•		•									Yes	Yes
	•	•	•					•	•	•	•	•		•									No	No
•												•											No	No
•	•	•						•	•	•	•	•		•									Yes	No
•	•		•	•	•	•	•		•	•	•	•		•		•	•						Yes	Yes
	•						•			•	•	•		•									Yes	No
•	•		•	•	•		•		•	•	•	•	•	•					•	•		•	Yes	No
•	•	•	•	•	•	•	•		•	•	•	•	•			•			•	•			Yes	No
•	•		•	•	•	•		•	•	•	•	•											No	No
	•				•			•		•	•	•		•									Yes	No
•	•			•	•	•	•	•			•												No	No

Table 2. System Functional Capabilities (continued)

System Name	Cartographic Output				Input Device Support						Output Device Support					
	Raster Maps	Vector Maps	3-Dimensional (Perspective) Plots	Georeferenced Overlay	Digitizing Tablet	Mouse	Light Pen	Touch Screen	Scanner Input	Other	Pen Plotter	Ink Jet Plotter	Electrostatic Plotter	Laser Printer	Dot Matrix Printer	Other
A-BASE	●	●	●	●	●	●	●	●	●		●	●	●	●	●	Numerous Graphic Cards
ACDS		●			●	●			●	Stereoplotter & Elec. Notebks	●		●	●	●	
ADDMAPS		●	●		●	●					●	●	●	●	●	
ARC/INFO	●	●	●	●	●	●	●		●		●	●	●	●	●	Film output devices for map publishing, cartography
ATLAS*DRAW					●	●					●	●		●	●	
ATLAS*GIS					●	●					●	●	●	●	●	
ATLAS*GRAPHICS											●	●	●	●	●	
ATLAS*MAPMAKER					●	●					●	●		●	●	
ATLAS/AGIS	●	●		●	●	●			●	ASCII DATA	●	●	●		●	Film Writers
ATOM	●		●	●	●	●			●	Digital tapes and film						Digital tapes and film
AUDRE	●	●			●	●			●		●	●	●	●	●	
Autoroute Plus	●					●								●	●	
CableCad				●	●	●					●	●	●	●	●	
CADCore/Hybrid	●	●			●	●			●		●		●			Thermal
CADKEY					●	●					●	●	●	●	●	
Cartographer		●				●			●		●	●		●	●	
CDM: GIM		●		●	●	●					●	●	●	●	●	
CDM: MM		●		●	●	●					●	●	●	●	●	
Census Windows		●				●	●				●			●	●	
CPS-3		●	●	●	●	●			●		●	●	●	●	●	
CRIES-GIS	●		●		●	●						●			●	
DIGI-CAD		●	●		●	●					●	●	●	●	●	
EARTH ONE	●	●	●		●	●			●		●	●	●	●	●	
EARTH/INFO		●			●	●					●	●			●	
EASI/PACE	●	●	●	●	●	●			●			●		●	●	Film Recorder, Thermal Wax Transfer
EIFM		●			●	●					●	●	●	●	●	
ERDAS	●	●	●	●	●	●	●		●	9-Track	●	●	●	●		Film recorder
FMS/AC		●	●	●	●	●			●		●	●	●	●	●	all supported by AutoCAD
GDS	●	●	●	●	●	●			●		●	●	●	●	●	
Genamap	●	●	●	●	●	●			●	total stations	●	●	●	●	●	Screen Copy Device
Geo/SQL		●	●	●	●	●	●	●	●		●	●	●	●	●	
Geographic Facilities Information System	●	●		●	●	●			●		●				●	
GRNDB		●		●							●	●	●			Graphics terminals
GIFM		●			●	●					●	●	●	●	●	
GIS CAD		●	●	●	●	●			●		●	●	●	●	●	
GIS Plus		●			●	●					●		●	●		
GIS/AMS		●		●	●	●			●		●	●	●	●	●	Screen copy device
GRASS	●	●	●	●	●	●					●	●	●	●	●	
GWN-GIS		●	●		●	●	●				●	●	●	●	●	
Chains	●	●		●	●	●	●			VideoImageCapture, Digitizing	●	●	●	●	●	Thermal
IDRISI	●	●	●		●	●			●			●		●	●	

Does System Provide User-generated Macros?	Remote Sensing Image Proc.: Geometric Corrections	Digital Image Classification	Classified Image Import to Database	Image overlay with GIS Map Data	Do You Plan to Support SDTS?
●					No
●					Yes
●					No
●	●		●	●	Yes
					No
●					No
●					No
					No
●	●	●	●	●	Yes
	●	●	●	●	Yes
●					Yes
					No
●					No
●					No
●					No
				●	Yes
●					Yes
					Yes
●					No
●					Yes
					Yes
					No
			●	●	No
●					Yes
●	●	●	●	●	Yes
●					No
●	●	●	●	●	Yes
●					Yes
					Yes
●					Yes
●					Yes
●				●	Yes
●					Yes
●					No
●					Yes
		●		●	Yes
●				●	Yes
					Yes
●	●		●	●	No
	●		●	●	No
	●	●	●	●	No

Table 2. System Functional Capabilities (continued)

System Name	Terrain Analysis with DEM Data								Miscellaneous										
	Compute Slope Ang	Determine Compass Aspect	Interpolate Elevation	Line of Sight Viewshed	Generate Elevation Contours	Model Drainage Network	Generate Terrain Cross-sections	Cut/Fill Calculations	Analysis Within Corridor	Boolean Operations-Multiple Maps	Boolean Operations-Multiple Themes	Nearest Neighbor Search	Moving Window Filt.	Comp. Optimal Path	Compute Weighted Optimal Path	COGO Computations	Network Analysis	Vector to Raster Conversion	Raster to Vector Conversion
INFO-MAP												●							●
INFORMAP	●	●	●	●	●	●	●	●	●	●	●	●		●		●	●		●
InSite*USA																			
KDMS					●		●												
LAND														●		●	●		
LandTrak									●	●	●	●		●	●		●		
MacGIS	●	●	●	●		●		●		●	●	●		●					
MacGIS																			
MACMAP									●		●		●	●	●	●	●	●	
Manatron											●								
MIPS	●	●	●		●	●	●		●	●	●	●	●					●	●
MapBox/Dragon	●	●	●	●		●		●	●	●	●	●	●	●	●		●		
MapGrafix			●		●		●	●	●	●	●					●			
MAPII	●	●	●	●		●	●		●	●	●	●	●	●	●		●	●	
MapInfo									●			●						●	
MapPower											●					●		●	
MARS																			
MGE	●	●	●	●	●		●	●	●	●	●	●		●	●	●	●	●	●
MIMS																●			
MOSS	●	●	●	●	●		●		●	●	●	●	●	●		●		●	
Nucor									●	●	●			●				●	●
PAMAP GIS	●	●	●	●	●	●	●	●	●	●	●	●	●	●	●	●	●	●	●
PMAP	●	●	●	●	●	●	●	●	●	●		●	●	●	●			●	
QUIKMap																			
SPANS	●	●	●	●	●	●	●	●	●	●	●	●	●	●	●		●	●	●
STRINGS			●	●	●		●		●	●	●					●			
Survey System		●			●		●	●								●			
System 600	●	●			●				●	●			●	●	●			●	●
System 9	●		●		●		●	●	●	●	●	●				●			
Terra-Mar GIS	●	●	●	●	●	●	●	●	●	●	●	●	●	●				●	●
Terrasoft	●	●	●	●	●	●	●	●	●		●	●	●			●		●	●
TIGRIS	●	●	●		●		●	●	●	●	●	●						●	●
Topologic	●	●	●	●					●	●		●						●	●
UltiMap	●	●	●	●	●		●	●	●	●	●	●	●	●	●	●	●		
USEMAP 4.2	●	●					●	●	●	●			●					●	●
Vortext	●	●	●	●	●	●	●		●	●	●	●	●			●		●	●
WIFM																●	●		
Windows/World														●				●	●
XMap	●	●	●	●	●		●		●			●		●			●		
No System Name	●	●	●	●	●	●	●	●					●			●	●	●	●
No System Name																			
No System Name																		●	●
No System Name																		●	●
No System Name	●		●		●		●	●								●			
No System Name	●	●	●		●		●	●											
No System Name	●			●	●	●	●		●			●	●					●	●
No System Name	●		●		●		●	●	●	●	●		●			●	●		
No System Name	●	●	●	●	●	●	●	●		●	●	●	●	●	●				●

User Interface and Display													Support of Standards											
Command Language	Menus	Icons	Multiple Windows	3-D Shaded Relief	3-D Wireframe	3-D Thematic Layer	Single Screen Oper.	Dual Screen Oper.	Multiuser Capability	Customizable Menu	Spatial Query by Cursor Input	Spatial Query by Coordinate Input	X-Windows	SQL	IRDS (FIPS 156)	GKS (FIPS 120)	PHIGS (FIPS 153)	CGM (FIPS 128)	POSIX (FIPS 151-1)	TCP/IP	GOSIP (FIPS 146-1)	ISO Network Stds.	Does System Provide On-line Help	Is On-line Help Context-sensitive
	●		●				●				●	●											No	No
●	●	●	●				●		●	●	●	●	●	●						●			Yes	Yes
●	●		●				●		●	●												●	Yes	Yes
	●							●	●											●			Yes	No
●	●							●	●	●	●	●		●						●			Yes	Yes
	●							●		●	●	●											No	No
●	●	●			●		●				●	●											Yes	No
	●	●									●												No	No
	●	●						●			●	●											Yes	Yes
	●		●				●		●	●	●	●											Yes	Yes
	●	●	●	●	●	●		●		●	●	●											Yes	Yes
●	●			●		●	●				●												Yes	Yes
	●	●	●				●	●		●	●	●		●						●			No	No
●	●	●	●				●	●			●												No	No
●	●	●					●		●	●	●	●						●					Yes	Yes
	●	●	●				●		●	●	●	●		●									Yes	Yes
	●	●	●				●	●	●		●	●											Yes	Yes
●	●	●	●	●	●	●	●	●	●	●	●	●	●	●		●			●	●		●	Yes	Yes
	●						●	●	●		●	●											No	No
●	●		●		●	●	●	●	●	●	●	●		●									Yes	No
	●	●					●	●			●												Yes	No
●	●	●	●	●	●	●	●	●	●	●	●	●	●	●			●	●	●	●	●	●	Yes	Yes
●	●											●											Yes	Yes
	●		●				●			●													Yes	No
●	●	●	●	●	●	●	●	●	●	●	●	●	●	●				●	●	●			Yes	Yes
	●				●	●	●	●	●	●	●	●		●									Yes	Yes
	●						●		●		●	●											No	No
●	●	●	●	●	●	●		●	●														Yes	Yes
●	●	●	●	●	●		●		●	●	●	●		●						●		●	Yes	Yes
●	●		●	●	●	●	●	●	●	●	●	●											Yes	No
●	●				●	●	●	●	●	●	●	●											Yes	Yes
●	●	●	●	●	●	●	●	●		●	●	●	●			●				●		●	Yes	Yes
●	●		●	●	●		●			●	●	●				●							Yes	No
●	●	●	●				●		●	●	●	●		●								●	Yes	Yes
●	●						●				●	●											Yes	Yes
●	●	●	●	●	●		●		●	●	●	●	●			●		●		●			Yes	Yes
●	●	●						●	●	●	●	●		●									Yes	No
●	●	●	●				●		●	●	●	●											Yes	No
	●	●	●		●	●	●		●		●	●	●	●									Yes	No
●	●		●	●	●	●		●	●	●	●	●											Yes	No
●	●		●		●		●	●	●	●													Yes	No
●	●		●	●				●	●			●										●	Yes	Yes
	●						●	●					●	●				●		●			Yes	No
	●	●	●	●	●		●		●		●	●											Yes	No
●							●	●	●	●													Yes	No
●				●	●	●	●	●	●	●	●	●				●						●	Yes	Yes
●	●	●	●	●	●	●	●	●		●	●	●	●	●						●			Yes	No
●	●	●	●	●		●	●	●	●		●		●	●						●		●	No	No

Table 2. System Functional Capabilities

System Name	ARC	AVHRR	DEM	DIME	DLG	DTED	DXF	ELAS	EPSF	ERDAS	ETAK	GIRAS	HPGL	IGES	ISIF	LANDSAT	MOSS	NTF	PICT	SIF	SPOT	TIFF	TIGER	Other
	Input Data Formats																							
INFO-MAP																								ASCII, DXF
INFORMAP				●	●		●				●			●			●			●			●	
InSite*USA																							●	
KDMS	●				●		●								●		●			●				
LAND				●	●										●					●			●	
LandTrak				●	●		●				●												●	
MacGIS			●																					
MacGIS																								
MACMAP																								Exel text or worksheet
Manatron			●	●	●		●								●					●			●	
MIPS	●	●	●		●		●	●		●			●			●	●				●	●	●	TARGA, RVF, GRASS
MapBox/Dragon		●	●				●			●						●								
MapGrafix					●		●				●												●	WDBII
MAPII:			●													●			●		●	●		SYLK
MapInfo				●			●																●	DBF, ASCII
MapPower	●			●	●		●				●		●	●		●			●		●	●	●	
MARS					●		●						●									●	●	
MGE			●	●	●	●	●				●	●	●	●	●	●	●			●	●		●	DFADS, GRASS, DGN
MIMS					●		●																●	ASCII coordinate lists
MOSS	●		●	●	●	●				●				●		●	●							GRASS, AMS, ADS
Nucor							●									●					●	●		
PAMAP GIS					●	●	●			●			●							●				IGDS
pMAP																								ASCII
QUIKMap							●																	IGDS, DGN, CARIS, NTX
SPANS	●	●	●	●	●	●	●	●		●	●		●		●	●	●			●	●		●	GINA, SPANS, IFF
STRINGS			●	●	●		●								●					●			●	
Survey System							●																	ASCII
System 600	●	●	●		●	●	●									●					●			DGN, GINA
System 9				●	●		●								●					●			●	CADDS 4X, WILD BC2
Terra-Mar GIS	●	●	●	●	●	●	●	●		●	●		●		●	●				●	●		●	
Terrasoft	●				●		●						●	●	●									COMARC
TIGRIS			●	●	●	●	●							●	●	●				●	●		●	DGN
Topologic	●						●																	TKP
UltiMap			●	●	●		●						●		●					●			●	
USEMAP 4.2																								
Vortext			●	●	●								●			●	●	●			●		●	not limited to above
WIFM				●	●															●			●	ASCII
Windows/World				●	●																			ASCII
XMap																								
No System Name	●			●	●		●			●				●	●					●			●	
No System Name																								
No System Name	●	●			●										●	●					●			
No System Name							●							●								●		CGM
No System Name																								AutoCad, Intergraph, others
No System Name			●				●							●	●					●				
No System Name	●		●	●	●	●	●	●								●				●				
No System Name	●		●	●	●		●				●			●									●	User Defined
No System Name	●		●			●				●						●					●	●		

Export Formats	Measurements				Generate Buffers					Map Algebra Functions					Polygon Operations					
	Straight Distance	Curved Distance	Proximity Analysis	Area Measurement	Around Points	Around Polygons	Along Straight Lines	Curved Lines	Weighted	Add/Subtract Maps	Mult./Divide Maps	Exponentiate Maps	Trig. Functions	Differentiate Values	Polygon Overlay	Point in Polygon	Line in Ploygon	Merge/Dissolve	Delete Spurious Polygons Overlay	Thiessen Polygon Generation
ASCII, DXF, PLX, TIFF, IMGE	●		●	●						●					●	●				
DXF, SIF	●	●	●	●	●	●	●	●	●	●	●	●	●	●	●	●	●	●	●	●
	●			●												●				
DXF, DLG, SIF, IGDS, MOSS	●			●																
IFF, open format graphics analysis	●		●	●	●	●	●			●	●					●	●	●	●	
DIME, TIGER, HPGL, DXF	●	●		●	●										●	●				
	●	●	●	●	●	●	●	●	●	●	●	●	●	●						
MacPaint		●		●	●	●	●	●		●	●				●					
Exel	●		●	●	●	●	●		●						●	●	●	●	●	
DXF	●		●	●	●	●	●	●								●	●			
ARC, TIGER, DXF, DLG, MOSS, ERDAS, TIFF, TARGA, GRASS, HPGL, MICROPIPS	●	●	●	●	●	●	●	●	●	●	●	●	●	●	●	●	●	●	●	
			●	●	●	●	●	●	●	●	●	●	●	●						
CGDEF - ASCII Text format	●	●		●	●	●	●			●						●	●			
PICT, TIFF, BINARY, ASCII, SYLK	●	●	●	●	●				●	●	●	●								
ASCII, DXF, HPGL, CGM, POSTSCRIPT	●		●	●	●					●					●	●	●	●		
ARC, DIME, DLG, DXF, ETAK, HPGL, IGES, LANDSAT, PICT, SPOT, TIFF, TIGER	●	●	●	●						●					●	●	●	●		
DXF, TIGER, TIFF	●	●	●	●											●	●	●	●	●	●
DLG, TIGER, ISIF, IGES, DGN	●	●	●	●	●	●	●	●		●	●	●	●	●	●	●	●	●	●	
DXF, HPGL, ASCII	●	●		●											●	●	●			
GRASS, SAGIS, IDIMS, DLG	●	●	●	●	●	●	●	●	●	●	●	●	●	●	●	●	●	●	●	
DXF, TIFF, Nuoor	●			●	●		●	●		●	●	●		●	●					
DLG, DTED, DXF, ERDAS, HPGL, SIF, IGDS	●	●	●	●	●	●	●	●	●	●	●	●	●	●	●	●	●	●	●	
Several ASCII formats			●	●	●	●	●	●	●	●	●	●	●	●						
HPGL, DXF, HALO, PIC	●	●		●						●										
DLG, DXF, HPGL, SIF, SPANS, RASTER, IFF	●	●	●	●	●	●	●	●	●	●	●	●	●	●	●	●	●	●	●	●
DLG, SIF, DGN, DXF	●	●	●	●	●	●	●	●		●	●	●	●	●	●	●	●	●	●	●
DXF, ASCII	●	●	●	●									●							
ARC, DXF, DGN, GINA, Generic raster	●		●	●	●	●	●	●	●	●	●	●	●	●						
SIF, DXF	●	●	●	●	●	●	●	●	●	●	●	●	●	●	●	●	●	●		
DLG, ARC, DXF, SIF, ERDAS	●	●	●	●	●	●	●	●	●	●	●	●	●	●	●	●	●	●	●	
IGDS, ARCINFO, DXF, ISIF	●	●	●	●	●	●	●	●	●	●	●	●	●	●	●	●	●	●	●	
DLG, DGN	●	●	●	●	●	●	●	●		●	●	●	●	●	●	●	●	●	●	
BITMAP, ASCII, DBF, DXF, HPGL, ARC	●			●	●	●	●			●	●	●				●	●	●		
DXF, SIF, ISIF, UltiMap	●	●	●	●	●	●	●	●	●	●	●	●	●	●	●	●	●	●	●	●
	●	●	●	●	●	●	●	●	●	●	●				●	●				●
Any	●	●	●	●	●	●	●	●		●	●	●	●	●	●	●	●	●	●	●
IFF	●	●																		
ASCII, BitMap	●	●		●	●										●	●	●			
	●	●	●	●	●										●	●	●			
	●	●	●	●	●	●	●	●	●	●	●			●	●	●	●	●	●	
DXF, CSV , HPGL, ARPLAN	●	●	●	●																
DLG-3, ISIF, FIRE, SCITEX, FILES-II, LGSWG	●	●		●											●	●	●	●	●	
DXF, HPGL, CGM, IGES, ISIF, Microcadam	●	●		●																
ASCII, AutoCad, Intergraph	●	●	●	●	●								●		●					
DXF, ISIF, IGDS, ASCII, MP Files	●	●		●						●	●		●	●						
Same as imput formats	●	●		●						●	●				●	●	●	●	●	
User Defined	●	●	●	●											●	●	●	●		
TIFF, ERDAS	●			●											●	●	●	●	●	●

System Name	Cartographic Output				Input Device Support						Output Device Support					
	Raster Maps	Vector Maps	3-Dimensional (Perspective) Plots	Georeferenced Overlay	Digitizing Tablet	Mouse	Light Pen	Touch Screen	Scanner Input	Other	Pen Plotter	Ink Jet Plotter	Electrostatic Plotter	Laser Printer	Dot Matrix Printer	Other
INFO-MAP	●				●	●			●		●			●	●	Ink jet printer
INFORMAP	●	●		●	●	●			●		●	●	●	●	●	
InSite*USA		●			●	●					●	●	●	●	●	
KDMS		●			●	●		●			●		●			
LAND	●	●			●				●		●	●	●	●	●	
LandTrak		●			●	●					●	●	●	●	●	
MacGIS	●		●	●		●			●		●			●	●	
MacGIS	●	●			●	●									●	
MACMAP	●	●			●	●	●	●		Scanner by link to a picture	●	●	●	●	●	Color printer
Manatron		●		●	●		●				●			●	●	
MIPS	●	●	●	●	●	●			●	Video Capture	●	●	●	●	●	Solid ink jet printer
MapBox/Dragon	●		●			●			●							
MapGrafix		●	●		●	●	●	●			●	●	●	●	●	
MAPII	●				●	●			●					●	●	
MapInfo		●			●	●					●	●	●	●	●	Slide Makers
MapPower	●	●			●	●		●	●		●	●	●	●	●	
MARS	●	●		●	●	●			●		●	●	●	●	●	
MGE	●	●	●	●	●	●			●	StereoPlotter, Elec. Survey Inst.	●	●	●	●	●	Film rec., Color thermal hardcopy
MIMS		●		●	●	●					●			●	●	
MOSS	●	●	●	●	●	●			●		●	●	●	●	●	
Nucor	●	●			●	●			●		●	●	●	●	●	
PAMAP GIS	●	●	●	●	●	●			●		●	●			●	
PMAP	●	●	●		●									●	●	
QUIKMap	●	●		●	●	●	●				●	●	●	●	●	
SPANS	●	●	●	●	●	●			●		●	●	●	●	●	
STRINGS		●	●	●	●	●			●		●	●	●	●	●	
Survey System				●	●	●				Field Data Recorder	●		●	●	●	
System 600	●		●	●	●	●			●				●	●	●	
System 9		●	●	●	●	●				Intergrated photogrametric	●	●	●	●	●	
Terra-Mar GIS	●	●	●	●	●	●			●		●	●	●	●	●	Thermal color
Terrasoft	●	●	●	●	●	●				Stereoplotter	●	●	●	●	●	
TIGRIS	●	●	●	●	●	●			●		●	●	●	●	●	Film rec, Color thermal hardcopy
Topologic	●	●			●	●					●	●	●	●	●	
UltiMap	●	●		●	●	●			●		●	●	●	●	●	
USEMAP 4.2	●	●	●	●	●				●		●	●		●	●	
Vortext	●	●	●	●	●	●					●	●	●	●		
WIFM		●			●	●					●	●	●	●	●	
Windows/World	●	●			●	●	●		●		●			●	●	
XMap	●	●	●	●		●			●		●	●	●	●	●	
No System Name	●	●	●	●	●	●			●		●		●		●	
No System Name					●	●					●	●	●	●	●	
No System Name	●				●	●			●			●	●	●	●	Thermal Wax, Photographic
No System Name	●	●				●			●		●		●	●	●	
No System Name		●	●	●	●						●		●			
No System Name		●				●					●				●	
No System Name	●	●	●	●	●				●	Laser Digitizer	●	●	●			
No System Name		●	●	●	●	●		●	●		●	●	●	●	●	Thermal transfer
No System Name	●		●	●		●			●			●	●	●		Thermal, Video Copy, Film

Does System Provide User-generated Macros?	Remote Sensing Image Proc. Geometric Corrections	Digital Image Classification	Classified Image Import to Database	Image overlay with GIS Map Data	Do You Plan to Support SDTS?
					No
•				•	Yes
•					Yes
•					Yes
•	•				No
•					No
•				•	No
					No
•				•	Yes
					No
	•	•	•	•	Yes
•	•	•	•	•	Yes
					Yes
	•	•	•	•	No
					No
	•	•	•		No
•					Yes
•	•	•	•	•	Yes
					Yes
•			•		Yes
•	•	•	•	•	No
•			•	•	Yes
•					Yes
					No
•		•	•	•	Yes
•					No
•					Yes
•	•	•	•	•	Yes
•					Yes
•	•	•	•	•	Yes
•			•	•	Yes
•	•	•	•	•	Yes
•					Yes
•			•		Yes
•					No
•			•	•	Yes
•					No
					Yes
	•	•	•	•	Yes
•					No
•					No
•	•	•	•	•	Yes
				•	No
					No
					Yes
•	•	•	•	•	Yes
•			•	•	Yes
•	•	•		•	Yes

Appendix F

State Contacts with Groundwater Database Interests

U.S. ENVIRONMENTAL PROTECTION AGENCY

Region IV
Contact:
George Collins
U.S. EPA Region iv
345 Courtland St., NE
Atlanta, GA 30365

Region IX
Contact:
Mark Rathbun
Systems Analyst
U.S. EPA Region IX
215 Fremont St.
San Francisco, CA 94105

Region VIII
Contact:
Pam Herman
U.S. EPA Region VIII
Water Management Division
Ground-Water Branch
999 18th St.
Suite 500
Denver, CO 80202-2405

STATE DATABASE SYSTEMS

ALASKA

Contact:

James Munter
Department of Natural Resources
Division of Geological and Geophysical Surveys
Water Resources Section
P.O. Box 2116 (Fish Hatchery Road)
Eagle River, AK

ARIZONA

Contact:

Howard Stapleton
John Wegrzyn
Department of Water Resources
99 E. Virginia Ave.
Phoenix, AZ 85004

CONNECTICUT

Contact:

Michael P. Prisloe
Natural Resources Center
Connecticut Department of Environmental Protection
165 Capitol Ave.
Hartford, CT 06106

FLORIDA

Contact:

James E. McNeil
Administrator
UIC and Technical Support Section
Bureau of Ground Water Protection
Twin Towers Office Building
2600 Blair Stone Rd.
Tallahassee, FL 32399-2400

GEORGIA

Contact:

Barry L. Johnson
Department of Health and Public Service
Agency for Toxic Substances and Disease Registry
Atlanta, GA 30333

ILLINOIS
Contact:
Susan Schock
Groundwater Section
Illinois State Water Survey
615 E. Peabody
Champaign, IL 61820

Contact:
Richard Carlson, Director
Illinois Environmental Protection Agency
220 Churchill Rd.
Springfield, IL 62706

Contact:
Warren Brigham
Illinois Department of Energy & Natural Resources
State Natural History Survey Division
Natural Resources Building
607 E. Peabody Dr.
Champaign, IL 61820

INDIANA
Contact:
Martin R. Risch
State of Indiana
Department of Environmental Management
Ground Water Section
Office of Environmental Response
105 South Meridian St.
P.O. Box 6015
Indianapolis, IN 46206-6015

KANSAS
Contact:
Vic Robbins, P.E.
Environmental Engineer
Bureau of Water Protection
Forbes Field
Topeka, KS 66620-0001

KENTUCKY
Contact:
Myles Powers
Kentucky Natural Resources Information System

Natural Resources and Environmental Cabinet
14th Floor Capital Plaza Tower
Frankfort, KY

LOUISIANA

Contact:

George H. Cramer II
Louisiana Department of Environmental Quality
Ground Water Protection Division
P.O. Box 44307
Baton Rouge, LA 70804

MINNESOTA

Contact:

Carrie Younkers
Project Analyst
Land Management Information System
Room 65 Metro Square Building
St. Paul, MN 55101

NEBRASKA

Contact:

Mahendra K. Bansal
State of Nebraska
Natural Resource Commission
301 Centennial Mall South
P.O. Box 95876
Lincoln, NE 68509-4876

OHIO

Contact:

John Voytek
Ohio Department of Natural Resources
Division of Water
Fountain Square, Bldg. E-3
Columbus, OH 43224

Contact:

Karen S. Wilson
U.S. Geological Survey
Water Resources Division
375 West Third Ave.
Columbus, OH 43212

PENNSYLVANIA
Contact:
Nicholas Molina
Michael J. Haydon
Department of Environmental Resources
Bureau of Water Quality Management
P.O. Box 2063 Fulton Bank Building
Harrisburg, PA 17120

SOUTH DAKOTA
Contact:
Gale N. Selken
DWNR Office of Water Quality
Joe Foss Building
Room 416
Pierre, SD 57501

WASHINGTON
Contact:
David Wilson
U.S. Geological Survey
Water Resources Division
1201 Pacific Ave. #600
Tacoma, WA 48402

WISCONSIN
Contact:
Alan Lulloff
Department of Natural Resources
Wisconsin Groundwater Information
P.O. Box 7921
Madison, WI 53707-7921

INDEX